Paleomicrobiology

Didier Raoult • Michel Drancourt
Editors

Paleomicrobiology

Past Human Infections

Springer

Prof. Didier Raoult
Prof. Michel Drancourt

Unité des Rickettsies
CNRS UMR 6020
Faculté de Médecine
Université de la Méditerranée
27 Boulevard Jean Moulin
13385 Marseille Cedex 5
France

Cover Photo: Detail of a painting by Michel Serre featuring 1720 Marseilles' plague epidemics (Musée des Beaux-Arts, Marseilles, France)

ISBN: 978-3-540-75854-9 e-ISBN: 978-3-540-75855-6

Library of Congress Control Number: 2007939832

Cover design: WMXDesign GmbH, Heidelberg, Germany

Printed on acid-free paper

5 4 3 2 1 0

springer.com

Preface

Almost 15 years ago, initial reports of the molecular detection of *Mycobacterium tuberculosis* DNA in ancient human skeletons of individuals suspected of having tuberculosis launched paleomicrobiology as an emerging field of research at the intersection of microbiology and evolution, history and anthropology. Refinements in experimental protocols together with strict criteria for determining the authenticity of data now allow the molecular diagnosis of past infections such as plague, tuberculosis, leprosy, typhoid fever, bartonelloses and influenza. Pioneering studies have compared the genotypes of organisms responsible for infection in past centuries with modern strains in order to gain a better understanding of microbe evolution. Paleomicrobiology provides historians and anthropologists with demonstrative data with which to analyse mass burials and past epidemics and their impact on human populations. These data help to resolve controversies regarding the aetiology of past epidemics such as the Black Death. Continuing progress in analytical techniques may allow further diagnoses of epidemics of as yet unknown aetiology and increased insight into the epidemiology of past infections. Looking backwards to past epidemics using modern tools and concepts will in turn help to understand the continuous evolution of microbes and of their direct and indirect relationships with humans.

This book summaries, for the first time, the concepts and techniques used to explore past epidemics and infections, and serves to illustrate the fruitful dialogue between historians, anthropologists and microbiologists through selected examples of research in the field of paleomicrobiology.

September 2007

Didier Raoult
Michel Drancourt

Contents

Contributors

Aboudharam, Gérard
Unité des Rickettsies, CNRS UMR 6020, Faculté de Médecine, Université de la Méditerranée, 27 Boulevard Jean Moulin, 13385 Marseille Cedex 5, France, and Département d'Odontologie Conservatrice, Faculté d'Odontologie, Université de la Méditerranée, Marseille, France

Castex, Dominique
CNRS, UMR 5199 PACEA, Laboratory of Anthropology of Past Populations, Bordeaux 1 University, Avenue des Facultés, 33405 Talence cedex, France, d.castex@anthropologie.u-bordeaux1.fr

Cunha, Burke A.
Infectious Disease Division, Winthrop-University Hospital, Mineola, New York, and State University of New York School of Medicine, Stony Brook, New York, llusardi@winthrop.org

Cunha, Cheston B.
The Pennsylvania State University College of Medicine, Hershey, Pennsylvania, ccunha@hmc.psu.edu

Donoghue, Helen D.
Centre for Infectious Diseases and International Health, Department of Infection, Windeyer Institute of Medical Sciences, UCL, 46 Cleveland Street, London W1T 4JF, UK, h.donoghue@ucl.ac.uk

Drancourt, Michel
Unité des Rickettsies, CNRS UMR 6020, Faculté de Médecine, Université de la Méditerranée, 27 Boulevard Jean Moulin, 13385 Marseille Cedex 5, France, Michel.Drancourtt@medecine.univ-mrs.fr

Duday, Henri
CNRS, UMR 5199 PACEA, Laboratory of Anthropology of Past Populations, Bordeaux 1 University, Avenue des Facultés, 33405 Talence cedex, France, h.duday@anthropologie.u-bordeaux1.fr

Dutour, Olivier
Unité d'Anthropologie, UMR 6578 CNRS Université de la Méditerranée, Faculté de Médecine, 27 Boulevard Jean Moulin, 13385 Marseille Cedex 5, France, olivier.dutour@medecine.univ-mrs.fr
Present address: Department of Anthropology, University of Toronto, 19 Russel Street, Toronto, ON M5S 2S2, Canada, oj.dutour@utoronto.ca

La, Vu Dang
Unité des Rickettsies, CNRS UMR 6020, Faculté de Médecine, Université de la Méditerranée, 27 Boulevard Jean Moulin, 13385 Marseille Cedex 5, France, danglv2001@yahoo.com

Lepidi, Hubert
Unité des Rickettsies, CNRS UMR 6020, Faculté de Médecine, Université de la Méditerranée, 27 Boulevard Jean Moulin, 13385 Marseille Cedex 5, France, Hubert.Lepidi@medecine.univ-mrs.fr

Lina, Bruno
Laboratoire de Virologie Est, Institut de Microbiologie, Centre National de Référence des virus influenza région sud, CBPE, GHE, Boulevard Pinel, 69677 Bron cedex, France; Virologie et Pathogenèse Humaine, UCBL-CNRS FRE3011, Université de Lyon, Faculté de Médecine R.T.H. Laennec, rue G. Paradin, 69372 Lyon, cedex 08, France, bruno.lina@univ-lyon1.fr

Mumcuoglu, Kosta Y.
Department of Parasitology, Hebrew University-Hadassah Medical School, P.O.B. 12272, Jerusalem 91120, Israel, kostam@cc.huji.ac.il

Nerlich, Andreas G.
Institute of Pathology, Academic Hospital Munich-Bogenhausen, Englschalkinger Str. 77, 81925 Munich, Germany, Andreas.Nerlich@extern.lrz-muenchen.de

Papagrigorakis, Manolis J.
Department of Orthodontics, Dental School, University of Athens, 2 Thivon Str., 11527 Goudi, Athens, Greece, and Museum of Craniofacial Evolution and Dental History, Dental School, University of Athens, 2 Thivon Str., 11527 Goudi, Athens, Greece, manjpap@dent.uoa.gr

Raoult, Didier
Unité des Rickettsies, CNRS UMR 6020, Faculté de Médecine, Université de la Méditerranée, 27 Boulevard Jean Moulin, 13385 Marseille Cedex 5, France, Didier.Raoult@medecine.univ-mrs.fr

Synodinos, Philippos N.
Department of Orthodontics, Dental School, University of Athens, 2 Thivon Str., 11527 Goudi, Athens, Greece, fsynodin@otenet.gr

Yapijakis, Christos
Laboratory of Molecular Neurobiology, Department of Neurology, University of Athens Medical School, Eginition Hospital, 74 Vas. Sofias Ave., 11528, Athens, Greece, cyapijakis_ua_gr@yahoo.com

Zink, Albert R.
Institute of Pathology, Academic Hospital Munich-Bogenhausen, Munich, Germany, Albert.Zink@lrz.uni-muenchen.de

Abbreviations

aDNA	ancient DNA
AR	absolute risk
BSA	bovine serum albumin
clyA	cytolysin A
CSD	cat scratch disease
DR	Direct Repeat
Fr1	unique fraction 1
GMS	Grocott-Gomori methenamine silver
H&E	hematoxylin and eosin
HIV	Human Immunodeficiency Virus
HPLC	high performance liquid chromatography
HSV	Herpes simplex virus
INRAP	Institut National de Recherches Archéologiques Preventives
MIRU	mycobacterial interspersed repetitive units
MS	mass spectroscopy
MST	multiple spacers typing
MTB	*Mycobacterium tuberculosis*
MTC	*Mycobacterium tuberculosis* complex
narG	nitrate reductase 1
osmC	osmotically inducible protein C
PAS	periodic acid-Schiff
PCR	polymerase chain reaction
PFGE	pulsed-field gel electrophoresis
PTB	*N*-phenacylthiazolium bromide
RFLP	restriction fragment length polymorphism
RR	relative risk
RT-PCR	reverse-transcription-polymerase chain reaction
SARS	severe acute respiratory syndrome
SNP	single nucleotide polymorphism
TB	tuberculosis
UV	ultraviolet
VNTR	variable number tandem repeats
WWI	First World War

Chapter 1
Great Plagues of the Past and Remaining Questions

Cheston B. Cunha(✉) and Burke A. Cunha

Abstract Due to the difficulty of obtaining tissue samples from victims of the ancient plagues, it is not always possible to utilise palaeomicrobiology techniques to determine the etiology of ancient infection. Therefore, it is often necessary to utilise other means to arrive at a likely diagnosis. The most helpful of these is the literary description of the disease. While this is often the best evidence available, working with such documents can prove difficult. Three great plagues of the ancient world, the Plague of Athens, the Antonine Plague, and the Justiniac Plague are described in either Latin or ancient Greek. The difficulties encountered when translating any ancient foreign language are compounded by the fact that so many words in these languages have a variety of meanings. This chapter reviews the three great plagues of antiquity from a clinical perspective.

1.1 Overview

There are numerous historical accounts of epidemics in ancient times. These accounts of epidemics or plagues describe infectious diseases ravaging ancient populations. Extant accounts of ancient plagues are limited. Certainly, many epidemics and outbreaks occurred in the ancient world that were not recorded or, if they were recorded, have been lost through the ages (Martin and Martin-Granel 2006). In the accounts that have survived, there are inherent difficulties of description and interpretation (Major 1978; Procopius 1981; Thucydides 1919). The descriptive terms used by the ancients are either not those that are we are familiar with today or, more problematic, multiple interpretations of terms are used to describe the physical findings in afflicted patients, which results in various possibilities and different assumptions are made according to translator variability and interpretation. Nevertheless, until recently, the only approach with which to try to determine the etiologies of ancient plagues has been examination of the relatively few written accounts that have survived over time. These accounts

Cheston B. Cunha
Pennsylvania State University College of Medicine, Hershey, Pennsylvania, USA
E-mail: ccunha@hmc.psu.edu

D. Raoult and M. Drancourt (eds.), *Paleomicrobiology: Past Human Infections.*

are variable and sometimes conflicting, and are dependent upon translator interpretation of the languages in which they were first described. Using the historical approach, there are other interpretational problems. The observers or recorders of the descriptions of ancient epidemics varied in their observational and descriptive capabilities as well as in their knowledge of medical terms used at the time (Page 1953; Parry 1969). All of these confounding variables make it difficult to determine the exact cause of the various plagues that afflicted the ancients (Brothwell and Sandison 1967; Cartwright 1972; Shrewsbury 1950).

In the absence of scientific proof, the historical method remains the backbone of an analytical approach to the problem of ancient plagues. Ancient descriptions provide at least some information about locale, season, descriptions of the findings, extent of the epidemic, mortality, and clinical sequelae. Although subject to interpretational difficulties, much has been learned by applying the historical method to determine the cause of widespread pestilence in earlier epics (Bollet 1987; Cunha 2004b; Cunha and Cunha 2006). Currently, methods are available to determine the actual cause of ancient plagues, but these methods depend on intact DNA for analysis, and are methodology dependent (Gilbert et al. 2003; Drancourt and Raoult 2005).

At the present time, sophisticated analytical methods are available to analyse ancient DNA in preserved tissue samples, permitting accurate identification of microorganisms present in samples from ancient animal and human remains (Cooper and Poinar 2000). As with the historical approach, there are problems with DNA-dependent technologies. The first difficulty using the scientific approach is to find suitably preserved samples to analyse. Just as there are problems with historical interpretation, so there are also problems with palaeomicrobiology (Hofreiter et al. 2001; Pääbo 1989). Firstly, there are relatively few geographical areas that have climatic conditions suitable for the preservation of tissue specimens in a state amenable to DNA analysis (Zink et al. 2002). The most likely situations in which DNA is likely to be preserved are in the dryness of the desert, in desiccated mummies, or tissues preserved in ice/glaciers (Jackson et al. 1998). DNA of microbial organisms or parasites from such specimens is likely to be well preserved and to lend itself readily to palaeomicrobiologic analyses (Arriaza et al. 1995; Li et al. 1999, Meers 1985; Reid et al. 2000, Rollo et al 2000; Spencer and Howe 2004; Taubenberger et al. 1997; Tumpey et al. 2004; Zink et al. 2002). Unfortunately, many of the ancient plagues that we are aware of from historical records did not occur in areas with favourable climatic conditions that would lend themselves to the preservation of analysable DNA samples (Antia et al. 2003; Brothwell and Sandison 1967; Cockburn 1971; Hofreiter et al. 2001; Zink et al. 2002). The ancient plagues of Egypt occurred in dry areas, but were not preserved in mummified remains.

The plagues of ancient Rome and ancient Athens occurred in climatic conditions that may not yield suitable specimens for palaeomicrobiologic identification (Kiple 1993; McNeill 1976). There are further problems with DNA specimen analysis, which has a corollary in contemporary clinical infectious diseases. In infectious diseases, one of the most fundamental determinations is to differentiate colonisation from infection. Similarly, in palaeomicrobiology, the mere recovery of an organism from an ancient preserved specimen does not necessarily implicate a role for this organism as the pathogen responsible for the demise of the individual whose remains are being analysed. The recovery of *Salmonella typhi* in areas endemic for enteric fevers as well as malaria and a variety of other infectious diseases, does not necessarily imply that the organism was causally related to the patient's demise. For example, if a patient who has pulmonary tuberculosis dies of coronary artery disease, the presence of tuberculosis does not necessarily imply that tuberculosis was the cause of death. Nevertheless, the study of palaeomicrobiology has contributed greatly to our understanding of the ancient microbial milieu of humans and animals (Antia et al. 2003; Brothwell and Sandison 1967; Cockburn 1971; McNeill 1976).

The very fact that the presence of such organisms can be verified is of great scientific importance (Drancourt et al. 1998). The types of specimen that lend themselves most readily to analysis are those that are likely to survive the ravages of time, i.e. teeth, bone specimens, coproliths, etc. Palaeopathology, the study of pathological changes in ancient remains, has been very important in confirming the presence of various diseases in ancient times. Because palaeopathology depends upon observable pathological changes in ancient specimens, palaeopathology is most useful in identifying infectious diseases with observable pathological changes. Skeletal syphilis and tuberculosis are examples of infectious diseases that produce characteristic changes in bone, which are readily recognisable in palaeopathological specimens (Arriaza et al. 1995; Brothwell and Sandison 1968; Kiple 1993). Infectious diseases that kill rapidly leave no traces in teeth, bone, or coproliths, which is problematic. In the absence of permanently preserved specimens, how would scientists in the future determine the presence/lethality of severe acute respiratory syndrome (SARS) in tissue from cadaveric specimens from Asia? Palaeomicrobiology has been most successful in demonstrating bacteria, Rickettsia, and parasite ova (Drancourt and Raoult 2005).

Palaeopathology has also demonstrated non-microscopic parasites in tissue specimens. Such findings are interesting and add to our knowledge of the epidemiology of infectious diseases in the ancient world (Arriaza et al. 1995; Brothwell and Sandison 1967). Although epidemiological analyses provide the background for the endemic illnesses in ancient populations, they do not explain the causes of the various plagues described in the ancient world (Kiple 1993). Until there is incontrovertible proof based upon methodologically sound science, the best current and future approach of trying to determine the etiology of epidemics in the ancient world is to combine the epidemiological information from palaeopathology with the continuing advances made in palaeomicrobiology, and this information should be used in conjunction with historical analyses (Cunha 2004b; Cunha and Cunha 2006). This chapter will take a historical approach combined with what is currently known from palaeomicrobiological and palaeopathological information to review the likely causes of three key ancient plagues of the past.

1.2 Determination of the Cause of Ancient Plagues by Historical/Clinical Analysis

While palaeopathology is probably the best way to obtain a definitive diagnosis of an ancient disease, it is not always possible. Due to the difficulty of obtaining tissue samples from victims of the ancient plagues, it is not always possible to utilise the technique of palaeomicrobiology to determine the etiology of ancient infection (Cooper and Poinar 2000; Drancourt and Raoult 2005; Hofreiter et al. 2001; Zink et al. 2002). Therefore, it is often necessary to utilise other means to arrive at a likely diagnosis. The most helpful of these is a primary literary description of the disease. While this is often the best evidence available, working with these sorts of documents can prove difficult. When it comes to discussing the three great plagues of the ancient world, the Plague of Athens, the Antonine Plague, and the Justiniac Plague, the descriptions are in either Latin or ancient Greek.

Therefore, The difficulties encountered when translating any ancient foreign language are compounded by the fact that so many words in these languages have a variety of meanings. Additionally, due to the precision required in medical documentation, any word or phrase that is interpreted in a way other than that intended by the original author can skew a description toward or away from the actual diagnosis (Cunha 2004b; Littman and Littman 1973; Major 1978; Procopius 1981). While Thucydides' chronicle of the plague is exquisitely detailed, variation in translation makes it impossible to definitively determine the causative agent (Page 1953; Parry 1969; Shrewsbury 1950; Thucydides 1919). Because of this, the only way to confirm a suspected diagnosis would be through the use of palaeomicrobiology (Drancourt and Raoult 2005). Indeed, a mass grave immediately outside Athens has been unearthed, but has not yet been analysed – until it is, debate will continue.

There is also an underlying assumption that the description was accurate to begin with and has been preserved intact (Major 1978). In the case of the Antonine plague, mere fragments of Galen's writings describing the course of the disease remain. Enough of the text is available to develop a clinical diagnosis, but this will need to be confirmed by palaeomicrobiological testing (Drancourt and Raoult 2005).

Obviously then, the best way to determine the cause of an ancient disease would be to combine palaeopathology with a literary clinical/historical analysis. This is the case with the Justiniac plague, which not only has a clear description that leads the reader to only one obvious conclusion, but also has evidence from mass graves from the era of the plague. These have been unearthed and genetic testing has confirmed the suspected etiology (Drancourt and Raoult 2002, 2004; Drancourt et al. 2004).

1.2.1 The Plague of Athens (430–426 B.C.): Determination of Etiology by Historical/Clinical Analysis

1.2.1.1 Historical Overview

Without doubt, Athens and Sparta were the two most powerful and influential civilisations on mainland Greece in the ancient world. By 431 B.C. the Peloponnesian War between Athens and Sparta had begun in earnest. Athens had only to survive the Spartan assault in order to claim victory, while the Spartans would have to conquer the city of Athens itself. Pericles, the leader of Athens, realised this and called for the Athenians to surrender their territory in Attica and to move all people in

Athens, and in the regions immediately surrounding it, into the city itself, which was protected by the great Themistoclean walls. These walls guarded the city proper, and provided a fortified connection with the harbour of Piraeus, 9 km from the city. Taking into account the Athenian's well-established naval superiority as well as their safe access to a protected port, it seemed as though taking Athens would be next to impossible for the Spartans. However, by 404 B.C., several events occurred that resulted in the total defeat of Athens and her allies. Most significant of these is the great Plague of Athens, described so accurately by Thucydides, the Greek historian. The plague struck Athens early in the conflict, during the summer of 430 B.C., and drastically reduced the population of the city, devastating Athenian society (Bollet 1987; Brothwell and Sandison 1967; Kiple 1993).

There has been much debate by both physicians and classicists as to the exact cause of the plague and neither group has come to a consensus. Although Thucydides was not a trained physician, he was most certainly an astute observer, and was careful to utilise the medical vocabulary of his era. Thucydides himself contracted and survived the plague, thus granting modern interpreters a precise and detailed account of the disease.

1.2.1.2 Thucydides' Clinical Description

It first began, so it is said, in ***Ethiopia above Egypt****, and then descended into Egypt and Libya and into most of the King's land. Suddenly falling upon Athens, it first attacked the population at Piraeus, so that they themselves said that the Peloponnesians had thrown poison into their cisterns: for there were, as yet, no wells there. But afterwards it came to the upper city as well, and from that time the deaths became much greater. Now, anyone, either physician or layman, can, by his own opinion, speak on its origins and the causes that produced so great a departure from normal conditions; but I shall talk about its course, and explain the symptoms, by which it could be recognised in the future, having knowledge of it beforehand. For I myself was ill and saw others suffer from it.*

That year, as agreed by all, had been unprecedentedly disease-free in respect to other sicknesses; but if anyone was suffering from anything at all before, all resolved into this. In other cases, there was no apparent cause, but suddenly, healthy men were seized first with ***mighty fevers in the head****, and* ***redness, and inflamed eyes,*** *and the inside, both the* ***throat and tongue****,* ***immediately became blood-red*** *and* ***emitted an atypical, foul breath****. After which came* ***sneezing and hoarseness****, and in not much time the pain descended into the chest, and produced a* ***severe cough****; and* ***when it fixed in the stomach, it upset it****, and* ***vomiting of bile*** *of every kind named by physicians ensued, accompanied by great suffering. In most cases* ***nonproductive retching*** *followed, giving way to* ***violent spasms****, which lessened, in some sooner, in others, not until much later. Externally, the body was* ***not very hot to the touch,*** *and was not pale,* ***rather, it was reddened, livid, and flowering with small blisters and wounds****. But* ***their insides burned so hotly, that the patients could not bare garments or fine cloths being laid on them, nor be anything but***

naked, and would have liked best to hurl themselves into cold water*, as in fact, many of those neglected did, throwing themselves into cisterns, tormented by unquenchable thirst. And it was the same whether they drank much or little. Also, they were ceaselessly tormented by the* ***inability to rest or sleep****. And the body, while the disease flourished, did not wither, but, contrary to expectations, withstood the ravages of the disease; so* ***that when they died, as most did, on the seventh or ninth day from the burning heat, they still had some strength****. But if they escaped this,* ***the disease descended into the bowels, resulting in a great ulceration****, and at the same time,* ***acute diarrhoea****. And many later* ***died from exhaustion*** *because of this.* ***For the disease, ran from above, in the head, where it first settled, throughout the whole body****, and if one survived the worst,* ***it left its mark on the extremities****.* ***For it fell upon the genitals, and the tips of the hands and the feet, and even having lost these parts, many survived,. Some also lost their eyes****. Others again were taken with a* ***complete loss of memory*** *after recovery, and they failed to know either themselves or friends.*

[Translation Thucydides (1919), bold/italics by Cheston B. Cunha].

1.2.1.3 Clinical Diagnostic Analysis

For over 2,000 years, physicians of every era have attempted to analyse Thucydides' writings and deduce the precise etiology of the Plague of Athens. Among the most likely diseases to have caused the plague of Athens are bubonic plague, typhoid fever, smallpox, measles, and epidemic typhus. All of these diseases were endemic to the ancient world and potentially fit the symptoms described by Thucydides (Longrigg 1980; McNeill 1976; Roberts and Manchester 2005; Shrewsbury 1950; Thucydides 1989). Unfortunately, the limitations presented by translation of the original Greek preclude a facile diagnosis, although it is possible to provide a satisfactory theory on the cause of the plague based upon the evidence provided by Thucydides. However, when carefully analysed, there is one disease that seems to fit the vast majority of symptoms more than the others (Kiple 1993; Osler 1876a, 1876b).

In modern times, whenever the word plague is mentioned to describe a disease, the most common thought is, of course, bubonic plague (Antia et al. 2003; Brothwell and Sandison 1967; Cockburn 1971). Boccaccio and others have passed down accurate descriptions of outbreaks of bubonic plague that occurred in Europe, and each resembles its fellows, but none of them bear a great likeness to Thucydides' account of the Plague of Athens. Indeed, the only symptoms of bubonic plague present in Athens during the plague years were fever and runny nose. Only if the Greek '*φλυκταιναις μικραις*' and '*ελκος*', meaning 'small blisters' and 'wounds', respectively, are interpreted as buboes, which is a linguistic stretch, would bubonic plague seem at all possible. However, none of the other features mentioned by Thucydides occur in bubonic plague outbreaks, suggesting that bubonic plague is a very unlikely cause of the Plague of Athens (Cunha 2004b; Page 1953; Shrewsbury 1950).

Typhoid fever is also an unlikely candidate. While the fever and diarrhoea are highly suggestive of typhoid, they are the only major symptoms that would indicate typhoid as the source of the plague. Typhoid fever requires fecal contamination of the water or food supply with *Salmonella typhi*, which could most definitely have occurred in the cramped, overpopulated conditions of wartime Athens. Even the rash, which is characterised by the presence of "rose spots", does not fit well with Thucydides' described rash. Additionally, typhoid usually causes death after 2–3 weeks, much longer than described by Thucydides. Also, typhoid fever does not confer complete immunity, while the Plague of Athens offered complete immunity in survivors. Finally, typhoid fever can be fatal, but usually does not approach a 25% mortality rate. It is for these epidemiological and clinical reasons that typhoid fever does not appear to be the cause of the Plague of Athens (Cunha 2004b; Cunha and Cunha 2006; Kiple 1993).

Smallpox is one of the diseased theorised to have caused the plague. This particular hypothesis, while put forward by many physicians over the years, was first suggested by the Persian physician Rhazes in 900 A.D. Thucydides' description has many features typical of smallpox. In particular, the rapid onset, fever, rash, and inflamed eyes all point to smallpox. Occurring in many forms, conventional smallpox or hemorrhagic smallpox is believed by some to be the most likely cause of the plague. Those who support the conventional smallpox hypothesis believe that the small blisters and wounds are indicative of the highly characteristic vesicles of smallpox. However, smallpox vesicles first appear as macules at the hairline and progress down from the face to the trunk, and that does not seem to be the rash that Thucydides describes. "Internal heat" has been mentioned in some smallpox cases, as has loss of vision, and gangrene of the extremities, but many other prominent symptoms are lacking. If hemorrhagic smallpox, the most lethal form of the disease, was the cause, then the vesicles of conventional smallpox would not develop, but rather a general petechial or purpuric rash, that is to say a rash characterised by small spots resulting from subcutaneous hemorrhage, would appear. Hemorrhagic smallpox bears a greater resemblance to Thucydides' description than the conventional form and would seem to be a very likely candidate for the cause of the plague. However, hemorrhagic smallpox never occurs independently of a conventional smallpox outbreak and, as such, Thucydides' description does not seem to be a description of a smallpox outbreak in Athens (Oldstone 1998; Osler 1876a, 1876b). In addition, the distribution of the rash and the lack of other symptoms all argue against smallpox being responsible for the plague of Athens (Aufderheide and Rodriguez-Martin 1998; Cunha 2002,; Cunha 2004a, 2004b; Fenner et al. 1988; Hopkins 2002; Osler 1876a, 1876b, 1892; Ricketts 1908).

One of the more likely causes, although one that is often overlooked due to the modern conception of the disease, is measles. When thinking of measles today, most conjure up the idea of a childhood disease, made less virulent by vaccination programs, incapable of inflicting the sort of damage witnessed in Athens in 430 B.C. However, when introduced to nonimmune populations, as happened on the Fiji Islands in 1875, mortality rates in all age groups approach the numbers described by Thucydides. Supporting measles as the cause of the plague are the rash, respiratory symptoms, restlessness, and "internal heat". Indeed, victims of the Fiji epidemic were often reported to throw themselves into rivers to find respite from the sensation of intense internal heat, strikingly similar to what the Athenians did in an attempt to alleviate their unbearable "inner heat". All this would seem to lead to a diagnosis of measles as the cause. However, measles rarely, if ever, presents with diarrhoea or gangrene, and the fact that Thucydides mentions these suggests they were present in most cases, rather than being rare occurrences among the sick. Moreover, neurological complications are rare in measles, which further suggests that measles did not, in fact, afflict Athens during the Peloponnesian War (Brothwell and Sandison 1967; Cartwright 1972; Cunha 2004a, 2004b).

Thucydides' description suggests that epidemic typhus was very likely the disease that ravaged Athens. Carried by lice, typhus has historically struck during times of war when a large population is forced to live in a relatively confined space, something that certainly would characterise Athens during the Peloponnesian War. This would allow a lice-infested individual to enter Athens through the port of Piraeus, which was the lifeline of Athens, and infect the entire city, killing close to one-quarter of the population. Considering the wartime conditions in Athens, which undoubtedly eliminated the typically excellent hygiene of Athenians, replacing it with the unsanitary habits of those who dwelled outside the city, it is very likely that epidemic typhus could have quickly spread throughout the population. Typhus is typically characterised by fever, red eyes, a truncal rash, and respiratory symptoms. Neurological complications are common, and may have been responsible for the blindness and the memory loss described. However, the most compelling arguments supporting epidemic typhus as the cause are the diarrhoea and gangrene, both of which are common in epidemic typhus. Finally, exhaustion is characteristic of those who die of this disease, and this is a feature clearly described by Thucydides.

By analysing the clinical features described by Thucydides, it seems that it was in fact epidemic typhus that caused the great Plague of Athens (McNeill 1976; Osler 1892; Shrewsbury 1950; Tumpey et al. 2004; Table 1.1). However, a definite etiology cannot be determined with certainty by clinical/historical means alone. The difficulties in using the historical approach are best illustrated using the Plague of Athens as a prime example (Christie 1969; Cunha 2004b; Cunha and Cunha 2006; Page 1953; Shrewsbury 1953).

1.2.1.4 Historical Importance

The plague of Athens had many direct and indirect consequences on the ancient Greek world, the most obvious of which was the depletion of manpower in Athens, i.e. by the winter of 427 B.C., Athens' fighting forces had been reduced to approximately 75% of their original strength. Indeed, virtually the entire eastern section of the Peloponnese lost close to 25% of its population. The plague did not, however, travel far enough to affect the Spartans and most of their allies, leaving Athens' foes virtually untouched by disease (Bollet 1987; Cartwright 1972).

Table 1.1 Differential diagnosis of athenian plague (Adapted from Cunha 2004b and Cunha and Cunha 2006)

Clinical description by Thucydides	Time of appearance	Bubonic plague	Typhoid fever	Smallpox	Measles	Epidemic typhus[a]
Rapid onset	Early	•		•	•	•
Fever	Early	•	•	•	•	•
Red eyes	Early			•	•	•
Runny nose and sneezing	Early	•			•	•
Red throat and hoarseness	Early				•	•
Foul breath	Early				•	•
Retching and convulsions	Middle					•
Livid red rash	Middle	•		•	•	•
Blisters and sores	Middle				•	•
Sensation of "intense internal heat"	Middle			•	•	•
Insomnia	Late	•		•		
Diarrhoea	Late		•		•	•
Tracheal/laryngeal ulcers	Late					•
Red throat and hoarseness	Late			•	•	•
Death by haemorrhage	Late					•

[a]Most likely etiology based on clinical/historical analysis

The plague also had a significant impact on Athenian political leadership, most notably the death of Pericles due to plague, which left Athens without one of its greatest statesmen. As a result of Pericles' death, subsequent Athenian leaders, such as Cleon, Alcibiades, and Hyperbolus, were allowed to shift Athens away from the noble course upon which it had been set by Pericles, and towards its eventual demise (Kiple 1993).

By depleting the ranks of the Athenian army, removing one of Athens' greatest leaders, and eliminating a system of beliefs and ideals that distinguished Athens from other ancient societies, the great Plague of Athens effectively altered the outcome of the Peloponnesian War, and subsequent Hellenistic and Western history (Cunha and Cunha 2006; Soupios 2004).

1.2.2 The Antonine Plague (166–270 A.D.): Determination of Etiology by Historical/Clinical Analysis

1.2.2.1 Historical Overview

By the second century A.D., the Roman Empire had asserted itself as, perhaps, the greatest civilisation the world has ever known. Rome was the preeminent force in the ancient world in terms of her cultural, political, economic, and military power, and Gibbon describes the reign of Marcus Aurelius, from 161–180 A.D., as "as the happiest and most prosperous period" in the history of humanity. This great empire stretched from the Iberian Peninsula to the Mid-East, and from Britain to North Africa, encompassing vast areas of the European continent.

When all of these grand achievements are viewed, it would seem as though a civilisation this splendid could never fall, and yet, by the third century, the Roman Empire was facing a period of crisis. This was in no small part due to the great plague that swept across the Empire in 166 A.D., which lasted for nearly a century. Later referred to as the Antonine Plague or the Plague of Galen, it decimated a large portion of the Empire's population and was a blow from which Rome never recovered (Bollet 1987; Brothwell and Sandison 1967; Cartwright 1972).

No great description exists for the Antonine Plague, unlike the earlier Plague of Athens. The only remaining documentation is in the form of several notes made by Galen, the great physician; an allusion to the epidemic by the emperor Marcus Aurelius in his writings, and two references by Lucian (Littman and Littman 1973; Major 1978). The plague originated in the Middle East and was brought to the Empire by Roman soldiers returning home after the Parthian War. Having thus been introduced into the Roman world, it spread rapidly, lasting until 270 A.D., claiming the lives of millions of Romans, including the Emperor Marcus Aurelius himself. Travelling via the Roman trade routes, there were even reports of the plague spreading as far east as China. So deadly was this pestilence that some sources suggest that, at one point during the plague years, over 2,000 people a day were dying in the city of Rome itself (Fears 2004; Gilliam 1961).

1.2.2.2 Galen's Clinical Description

Exanthem:

On the ninth day a certain young man was ***covered over his whole body with an exanthem****, as was the case with almost all who survived. Drying drugs were applied to his body.* ***On the twelfth day he was able to rise from bed.***

On those who would survive who had diarrhoea, ***a black exanthem*** *appeared on the whole body.* ***It was ulcerated in most cases and totally dry****. The blackness was due to a remnant of blood that had putrefied in the fever blisters, like some ash which nature had deposited on the skin. "Of some of these which had become* ***ulcerated, that part of the surface called the scab fell away and then the remaining part nearby was healthy and after one or two days became scarred over****. In those places where it was not ulcerated,* ***the exanthem was rough and scabby and fell away like some husk*** *and hence all became healthy. In many cases where there was* ***no bloody colliquescences*** *(diarrhoea), the entire body was covered by a black exanthem. "And sometimes a sort of scale fell off, when the exanthem had dried and dissipated, little by little, over a period of many days after the crisis.*

Fever:

Those afflicted with plague ***appear neither warm, nor burning to those who touch them, although they are raging with fever inside****, just as Thucydides describes.*

Galen calls the plague a fever plague.

Black excrement *was a symptom of those who had the disease, whether they survived or perished of it. Colliquescence (diarrhoea) was first auburn, the yellowish red, later black, like fecal matter of blood. Colliquescence of evacuation was an inseparable symptom of the plague. In many who survived, black stools appeared, mostly on the ninth day or even the seventh or eleventh day. Many differences occurred. Some had stools that were nearly black; some had neither pains in their excretions, nor were their excretions foul smelling. Very many stood in the middle. If the stool was not black, the exanthem always appeared.* ***All those who excreted very black stool died****.*

Vomiting:

Occurred in some cases.

Cough-Catarrh:

On the ninth day a young man had a ***slight cough****. On the tenth day the* ***cough became stronger and with it he brought up scabs****. After having catarrh for many days, first with a cough* ***he brought up a little bright, fresh blood****, and afterwards even part of the membrane which lines the artery and rises through the larynx to the pharynx and mouth.*

Internal Ulcerations and Inflammation:

On the tenth day a young man coughed and brought up a scab, which was an indication of an ulcerated area in the windpipe in the region of the trachea near the jugular vein. ***No ulcers were present in the mouth or throat*** *(there was no problem of ingesting food).* ***The larynx was infected, and the man's voice was damaged****.*

Duration of the Disease:

The crisis appeared on the ninth to twelfth day. ***On the third day after the ninth the young man was able to rise from his bed.***

[Translation from Galen by Littmann and Littmann (1973), bold/italics by Cheston B. Cunha].

1.2.2.3 Clinical Diagnostic Analysis

Although what has remained of Galen's description of the Antonine Plague is not as detailed as Thucydides' Athenian plague description, Galen's precise account of the exanthem that characterises the plague makes it relatively easy to pinpoint its cause. The fact that the rash extends over the entire body rather than being concentrated in the form of buboes in the groin and armpit rules out bubonic plague as a cause. Similarly, typhoid fever lacks most of the symptoms Galen describes, and thus is not a very likely cause of the plague (Christie 1969; Kiple 1993).

The exanthem has the potential to be the rash that is seen in measles, epidemic typhus or smallpox. Indeed, the early stages of these diseases are very easy to confuse. Undoubtedly, since measles, typhus, and smallpox have many of the same characteristics found in Galen's plague descriptions, such as the "internal heat", and foul breath, the type of vesicles must be looked to in order to differentiate between the three most likely possibilities. However, Galen's statement that the exanthem was pustular and later became blackened is highly suggestive of the pustular stage of a smallpox rash most often seen in hemorrhagic smallpox. Because of this evidence, smallpox seems to fit; however, there is one problem with that diagnosis (Kiple 1993; Tumpey et al. 2004; Table 1.2) Smallpox confers complete immunity, which would seem to work against the plague's recurrences later in the third century. Nevertheless, it is possible, and very likely, that these recurrences were merely instances of the plague entering previously unaffected areas of the Empire, making smallpox a perfectly viable cause. Therefore, after considering all the information regarding the symptoms of the diseases, Galen's description of a pustular rash indicates that the Antonine Plague was, in fact, an outbreak of smallpox (Littman and Littman 1973; McNeill 1976).

Table 1.2 Differential diagnosis of antonine plague

Clinical description by Galen	Time of Appearance	Bubonic plague	Typhoid fever	Measles	Epidemic typhus	Smallpox[a]
Rapid onset	Early	•		•	•	•
Fever	Early	•	•	•	•	•
Foul breath	Middle			•	•	•
Livid red rash	Middle			•	•	•
Blisters and sores	Middle				•	•
Sensation of "intense internal heat"	Middle			•	•	•
Insomnia	Late	•		•		
Diarrhoea	Late		•		•	•
Tracheal/laryngeal ulcers	Late					•
Red throat and hoarseness	Late			•	•	•
Death by haemorrhage	Late					•

[a]Most likely etiology based on clinical/historical analysis

It is clear that the scope of the plague was enormous, and impacted all levels of Roman life. Indeed, for an Empire so dependent on manpower for its financial, agricultural and military infrastructure, the plague was a crippling event from which the Empire never truly recovered. Ultimately, it can be said that the Antonine Plague had a profound negative impact on the spiritual, political, economic, social, and military aspects of the Roman Empire. Together with military defeats at the hands of the German tribes, the Antonine Plague most definitely set Rome on her long decline to ruin (Fears 2004; Gilliam 1961).

1.2.3 The Justiniac Plague (542–590 A.D.): Determination of Etiology by Historical/Clinical Analysis

1.2.3.1 Historical Overview

As bleak as things appeared to be for the Roman Empire at the end of the third century A.D., during the time between 290 A.D. and 540 A.D., the Roman Empire was able to regain a certain degree of vitality. While the power of the western Empire slowly waned, the eastern part of the Empire maintained its presence and, with it, the spirit of the Roman Empire. When Justinian took power in the east, he brought with him a dream of resurrecting the old might of Rome by reuniting the eastern and western portions of the Empire, and he was almost able to accomplish this. After securing his northern and eastern borders, Justinian began his campaign in the west in 532 A.D. Initially he met with great success, retaking much of Rome's lost territory. The new emperor had recaptured North Africa, Carthage, Sicily, parts of Hispania, and even large sections of the Italian peninsula. Indeed, by 540 A.D. German resistance was collapsing, and Justinian hoped to launch an attack into Gaul and possibly Britain as well. It seemed as though the reign of Justinian would be one of renewed glory and vigour for the Roman Empire, but all that changed when the Justiniac Plague struck in 542 A.D. (Brothwell and Sandison 1967).

Most likely carried from Africa, where it originated, to Constantinople and the rest of the Empire in a shipment of grain from Egypt, it was easy for the disease to spread along the trade routes of Rome. The enclosed city of Constantinople, as was the case with Athens during the Peloponnesian War, would have provided ideal conditions for the proliferation of the disease throughout the city's population (Bollet 1987).

There are several descriptions of the Justiniac Plague, namely those of John of Ephesus, Evagrius Scholasticus, and Procopius. While all provide descriptions, it is Procopius' account that is considered to be the most accurate and certainly the most precise in describing the symptoms of the plague. Procopius was one of the principal archivists for the Emperor Justinian, and had travelled for some time on campaign with Justinian's great general, Belisarius. When the plague arrived in Constantinople, where Procopius was staying, he chronicled his account of the plague (Kiple 1993; Major 1978).

1.2.3.2 Procopius' Clinical Description

During this time there was a plague, by which all men were almost completely killed…

… For it did not come in a certain part of the world or to certain men, nor did it confine itself to any season of the year, so that from such circumstances it might be possible to find explanations of a cause, but it encompassed the entire world, and destroyed the lives of all men, although they differed from one another in the most obvious ways, respecting neither sex nor age.

For just as men differ with regard to the places in which they live, or in the manner of their daily life, or in natural disposition, or in active endeavor, or in whatever else man differs from man, in the case of this disease alone, the difference meant nothing. And it attacked some in the summer, others in the winter, and still others at other times of the year. Now let each one express his own opinion concerning the matter, both sophist and astrologer, but as for me, I shall proceed to tell where this disease originated and the manner in which it destroyed men.

It came from the Egyptians who live in Pelusium*. But it split, and in one direction came toward Alexandria and the rest of Egypt, and in the other it came to Palestine bordering Egypt, and from there spread everywhere, always moving forward and going whenever time favoured it. For it seemed to move by a set plan and delayed in each land for a certain time, casting its blight leniently on none, but spreading in either direction right out to the ends of the world, as if afraid that some corner of the Earth might escape it. For it spared neither island nor cave nor mountain that had human inhabitants; and if it had passed over any land, either not affecting the men there or touching them in an inconsequential fashion, at a later time it still came back; then those who lived near this land, whom formerly it had most gravely afflicted, it did not touch at all, but it did not leave the place in question until it had given up its just and proper toll of dead, which corresponded exactly to the number killed at the earlier time among those who lived nearby. And this disease always* ***started on the coast and from there moved to the interior****. And in the second year it reached Byzantium in the midst of spring, where I happened to be staying at the time. And it came thusly. Many people saw the spirits of divine beings in human form of every kind, and, as it happened, those who encountered them thought that they were struck, in this or that part of the body, by the man they had met; and immediately seeing this apparition they were also seized by the disease. Now at first those who met these creatures tried to turn them aside by uttering the holiest of names and exorcising them in other ways as best each one could, but they accomplished absolutely nothing, for even in the sanctuaries, where the most of them fled for refuge, they were dying constantly. But later on they were unwilling to even listen to their friends when they called to them, and they shut themselves up in their rooms and pretended they did not hear, although their doors were being beaten down, fearing that he who was calling was one of those spirits. But in the case of some, the pestilence did not come in this way, but they saw a vision in a dream and seemed to suffer the very same thing at the hands of the creature who stood over them, or else to hear a voice prophesising that they were written down* *in the number of those who were to die. But with* ***most it*** *happened that the disease seized them* ***without*** *being made aware of what would come* ***by a waking vision or a dream****. And they were taken as follows.*

They had a ***sudden fever****, some when they woke from sleeping; others while walking around; and still others while busy with other matters, regardless of what they were doing. But the body showed no change in its original color, neither was it as hot as expected when struck by the fever, nor did any inflammation occur, but the fever was of such a lethargic kind from its onset until the evening that it would not give any suspicion of danger either to the sick themselves or to a physician. Therefore, it was natural that none of those who had contracted the disease expected to die because of it. But in some cases on the same day, in others on the following day, and in the rest, not many days later, a* ***bubonic swelling developed, there in the groin of body, which is below the abdomen, but also in the armpit, and also behind the ear and at different places along the thighs****. Up to this point, then, everything occurred the same way with all who had taken the disease. But from then on very distinct differences developed for there ensued for some a deep* ***coma****, with others a* ***violent delirium****, but, in either case, they suffered the characteristic symptoms of the disease. For those who were under the spell of the coma* ***forgot all those who were familiar to them,*** *and seemed to lie, sleeping constantly. And if anyone cared for them, they would eat without waking, but some also were neglected, and these would die directly through lack of sustenance. But those who were seized with delirium suffered from* ***insomnia*** *and were victims of a* ***distorted imagination****; for they suspected that men were coming to them to destroy them, and they would become excited and rush off in flight, crying out at the top of their voices. And those who were attending them were in a state of constant exhaustion and had a most difficult time. For this reason everybody pitied them no less than the sufferers, not because they were threatened by the pestilence by going near it, for neither physicians nor other persons were found to contract this plague through contact with the sick or with the dead, for many who were constantly engaged either in burying or in attending those in no way connected with them survived in the performance of this service beyond all expectation, while with many others the disease came on without warning and they died immediately; but they pitied them because of the great hardships which they were undergoing. For when the patients fell from their beds and lay rolling on the floor, they kept putting them back in place, and when they were struggling to rush headlong out of their houses, they would force them back by shoving and pulling against them. And when water happened to be nearby, they wished to fall into it, not so much because of a desire for drink, for the most of them rushed into the sea, but the cause was to be found chiefly in the diseased state of their minds.*

They also had great difficulty in the matter of eating*, for they could not easily take food. And many perished through lack of any man to care for them, for they were either overcome by hunger, or threw themselves from a height. And in those cases where neither coma nor delirium came on, the bubonic swelling became worse and the sufferer, no longer able to endure the pain, died. And one would suppose that in all cases the same thing would have been true, but since they did not* *all have their senses, some were unable to feel the pain; for owing to the troubled condition of their minds they lost all sense of feeling.*

In some cases death came immediately, in others, after many days; and with some the body broke out with ***black pustules*** *about as large as a lentil and these did not survive even one day, but all succumbed immediately.* ***Vomiting of blood*** *ensued in many, without visible cause, and immediately brought death. Moreover, I am able to declare this, that the most illustrious physicians predicted that many would die, who, shortly afterwards, unexpectedly escaped from suffering entirely, and physicians declared that many would be saved, who were destined to be carried off almost immediately. So it was that in this disease there was no cause that came within the realm of human understanding; for in all cases the issue tended to be something unaccountable.*

Now in those cases where the ***swelling rose to an unusual size and a discharge of pus had set in, it happened that they escaped from the disease and survived,*** *for clearly the acute condition of the swelling found relief in this direction, and this proved to be, in general, an indication of returning health; but in cases where the swelling maintained its former appearance, there ensued those troubles which I have just mentioned. And with some of them the thigh withered, in which case, though the swelling was there, it did not develop the least suppuration. With others who survived, the tongue did not remain unaffected, and they lived on either lisping or speaking incoherently and with difficulty.*

[Translation by H.B. Dewing from Procopius (1981), bold italics by Cheston B. Cunha].

1.2.3.3 Clinical Diagnostic Analysis

Unlike the Plague of Athens, after analysing the description of the symptoms and signs, the cause of the Justiniac Plague is apparent. The most notable of the symptoms described by Procopius is, of course, the bubonic swellings (*υσερον βουβων*), which developed in the groin and axilla of those who contracted the disease. Indeed, even without analysis of other signs and symptoms, this description is highly indicative of bubonic plague.

Perhaps the most unusual symptom described by Procopius was the visualisation of spirits that many infected individuals claimed to have seen. However, a common complication associated with a bubonic plague is encephalopathic. These visions may, in fact, be early manifestations of the neurological complications of the plague, i.e. encephalopathy, which may progress to coma and delirium in some. This type of neurological involvement is very different from epidemic typhus. Although typhus results in a loss of memory, hallucinations of the type described by Procopius occur only rarely. Similarly, while typhus presents with many of the features described by Procopius, it lacks many of the more critical symptoms, namely, the presence of buboes, and the development of coma (Christie 1969; Kiple 1993; Tumpey et al. 2004).

Similarly, measles also seems to lack the requisite symptoms and is unlikely to be the cause of the great Justiniac Plague. Both measles and typhoid fever show only a few basic features in common with the plague that descended on the Roman world in 542 A.D. The fever and diarrhoea of typhoid and the red throat of measles, when viewed by themselves, could be indicative of many infectious diseases, and do not provide enough commonality to warrant a diagnosis of either disease by description alone (Cunha 2004a, 2004b).

Finally, smallpox is a diagnostic possibility, but one that is not very likely when studied closely. As with the aforementioned diseases, the presence of buboes is not indicative of the type of rash seen in smallpox infections. Rather than the truncal, vesicular rash typical of variola, the swellings described concentrate in the primary lymphatic tissue of the groin, armpit, and neck (Christie 1969; Kiple 1993; Osler 1876a, 1876b, 1892). For this reason, along with the neurological and pulmonary complications often seen in plague, bubonic plague is almost certainly the cause of the Justiniac Plague (Bratton 1981a, 1981b; Tumpey et al. 2004; Table 1.3).

The damage resulting from the Justiniac Plague was both far reaching and disastrous for Rome. Although the precise numbers provided by Procopius and others who wrote about the plague are not always accurate, it can be safely assumed that well over one-third of the Roman world's population was eliminated by the conclusion of the sixth century A.D. Additionally, as Procopius describes, much of the surviving population of the Empire, who had become infected but did not perish, suffered from the debilitating and crippling neurological effects of the plague (Bollet 1987; Cartwright 1972).

Although Justinian sought to re-conquer much of Italy and the Western Empire, the plague effectively ended his plans of restoring much of what had been the old Roman Empire. These disastrous effects on the Roman Empire were compounded by the fact that the plague did not affect the less organised, "barbarian" societies

Table 1.3 Differential diagnosis of justiniac plague

Clinical description by Procopius	Time of appearance	Typhoid fever	Measles	Epidemic Typhus	Smallpox	Bubonic plague[a]
Rapid onset	Early		•	•	•	•
Slight fever	Early	•	•	•	•	•
Coma	Middle					•
Buboes	Middle					•
Delirium	Middle			•		•
Haematoemesis	Middle					•
Insomnia	Middle		•			•
Diarrhoea	Middle	•		•	•	
Red throat and hoarseness	Middle		•	•	•	•
Death by haemorrhage	Late				•	

[a]Most likely etiology based on clinical/historical analysis

outside of Rome's borders. This, in large part, was due to the fact that the highly developed internal structure of the Roman Empire actually facilitated the spread of bubonic plague along Rome's trade routes. The less centralised, foreign civilisations bordering the Roman Empire were, therefore, far less likely to have plague spread rapidly through their populations (McNeill 1976). It is, of course, impossible to claim that the eventual destruction of the Roman Empire was brought about solely by this plague, but it can be said that because of the Justiniac Plague, the Roman Empire lost any initiative it had recovered following the Antonine Plague. The plague so weakened the Eastern Roman Empire that it never truly recovered, and led to the eventual, complete collapse of the last remnants of the once mighty Roman Empire (Cunha and Cunha 2006; Kiple 1993).

1.3 Summary

The three great ancient plagues provide prime examples of the limitations and accuracy of clinical/historical analysis. Clinical historical analysis provides increasing diagnostic certainty going from the indeterminate certainty of the Plague of Athens, to more certainty with the Antonine Plague, and absolute certainty with the Justinian Plague. The challenges for the future are to find additional tissue samples that have sufficiently preserved microbiological DNA, which hopefully will provide definitive information on the cause of some of the ancient plagues (Drancourt and Raoult 2005). Mass burial sites that have previously remained uncovered may be found in the future, and analysis of such remains put in the correct clinical context can provide invaluable information regarding the infectious disease etiology of the causes of various ancient epidemics. Historical analysis will continue to be important because it provides the clinical descriptions of diseases, which may be the same as or different from current clinical descriptions of various infectious diseases (Brothwell and Sandison 1967; Kiple 1993). The historical approach is compounded by difficulties with infectious diseases that no longer exist, that occurred abruptly or more likely evolved over time, which created such mass devastation and then disappeared into the fog of history. Palaeomicrobiology is also of great importance in helping to sort out the evolution of infectious diseases on a microbe by microbe basis. A great breakthrough in palaeopathology has been the demonstration of microbial DNA in dental pulp specimens (Drancourt and Raoult 2002, 2004; Drancourt et al. 2005; Raoult et al. 2000). Bacteria early in bacteremia are trapped in dental pulp and preserved if the victim's teeth are preserved/found. To date, dental pulp DNA analysis is the only way to accurately identify microbial DNA from rapidly fatal infections of the past (Drancourt et al. 1998). Thus, although the future analyses of ancient plagues will still use the historical method as the foundation, as more well-preserved specimens are uncovered from ancient sites, and more DNA techniques are refined and standardised, ever more important information on the evolution of infectious disease agents as well as their role in ancient plagues will continue to interest and amaze us.

References

Antia R, Regoes RR, Koella JC, Bergstrom CT (2003) The role of evolution in the emergence of infectious diseases. Nature 426:658–661

Arriaza BT, Salo W, Aufderheide AC, Holcomb TA (1995) Pre-Columbian tuberculosis in northern Chile: molecular and skeletal evidence. Am J Phys Anthropol 98:37–45

Aufderheide AC, Rodriguez-Martin C (1998) Smallpox. In: The Cambridge encyclopedia of human palaeopathology. Cambridge University Press, Cambridge, pp 201–207

Bollet AJ (1987) Plagues and poxes. Demos, New York

Bratton TL (1981a) The identity of the Plague of Justinian. Trans Stud Coll Physicians Philadelphia 3:113–124

Bratton TL (1981b) The identity of the Plague of Justinian (Part II). Trans Stud Coll Physicians Philadelphia 3:174–180

Brothwell J, Sandison AT (1967) Diseases in antiquity. Thomas, Springfield, IL

Cartwright FF (1972) Disease and history: the black death. Dorset, New York

Christie AB (1969) Infectious diseases: epidemiology and clinical practice. Churchill-Livingstone, Edinburgh

Cockburn TA (1971) Infectious disease in ancient populations. Curr Anthropol 12:45–62

Cooper A, Poinar HN (2000) Ancient DNA: do it right or not at all. Science 289:1139

Cunha BA (2002) Smallpox: an Oslerian primer. Infect Dis Pract 26:141–148

Cunha BA (2004a) Smallpox and measles: historical aspects and clinical differentiation. Infect Dis Clin N Am 18:79–100

Cunha BA (2004b) The cause of the plague of Athens: plague, typhus, smallpox, or measles? Infect Dis Clin N Am 18:29–43

Cunha CB, Cunha BA (2006) Impact of plague on human history. Infect Dis Clin Am 20:253–272

Drancourt M, Raoult D (2002) Molecular insights into the history of plague. Microbes Infect 4:105–109

Drancourt M, Raoult D (2004) Molecular detection of *Yersinia pestis* in dental pulp. Microbiology 150:263–264

Drancourt M, Raoult D (2005) Paleomicrobiology: current issues and perspectives. Nature Rev Microbiol 3:23–35

Drancourt M, Aboudharam G, Signoli M, Dutour O, Raoult D (1998) Detection of 400-year-old *Yersinia pestis* DNA in human dental pulp: an approach to the diagnosis of ancient septicemia. Proc Natl Acad Sci USA 95:12637–12640

Drancourt M, Roux V, Dang LV, Tran-Hung L, Castex D, Chenal-Francisque V, Ogata H, Fournier P-E, Crubézy E, Raoult D (2004) Genotyping, Orientalis-like *Yersinia pestis*, and plague pandemic. Emerg Infect Dis 10:1585–1592

Fears JR (2004) The Plague under Marcus Aurelius and the decline and fall of the Roman Empire. Infect Dis Clin N Am 20:65–77

Fenner F, Henderson DA, Arita I (1988) Smallpox and its eradication. World Health Organization, Geneva

Gilbert MT, Willerslev E, Hansen AJ, Barnes I, Rudbeck L, Lynnerup N, Cooper A (2003) Distribution patterns of postmortem damage in human mitochondrial DNA. Am J Hum Genet 72:32–47

Gilliam JF (1961) The Plague under Marcus Aurelius. Am J Philol 30:225–251

Hofreiter M, Serre D, Poinar HN, Kuch M, Paabo S (2001) Ancient DNA. Nature Rev Genet 2:353–359

Hopkins DR (2002) The greatest killer: smallpox in history. University of Chicago Press, Chicago, pp 13–317

Jackson PJ, Hugh-Jones ME, Adair DM, Green G, Hill KK, Kuske CR, Grinberg LM, AbramovaFA, Keimi P (1998) PCR analysis of tissue samples from the 1979 Sverdovsk anthrax victims: the presence of multiple *Bacillus anthracis* strains in different victims. Proc Natl Acad Sci USA 95:1224–1229

Kiple KF (1993) The Cambridge world history of human diseases. Cambridge University Press, New York

Li HC, Fujiyoshi T, Lou H, Yashiki S, Sonoda S, Cartier L, Nunez L, Munoz I, Horai S, Tajima K (1999) The presence of ancient human T-cell lymphotropic virus type 1 provirus DNA in an Andean mummy. Nature Med 5:1428–1432

Littman RJ, Littman ML (1973) Galen and the Antonine plague. Am J Philol 94:243–255

Longrigg J (1980) The great plague of Athens. Hist Sci 18:209–225

Major RH (1978) Smallpox and measles. In: Classic descriptions of disease. Thomas, Springfield, IL, pp 196–200

Martin PMV, Martin-Granel E (2006) 2,500-year evolution of the term epidemic. Emerg Infect Dis 12:976–980

McNeill WH (1976) Plagues and peoples: a natural history of infectious diseases. Anchor, New York

Meers PD (1985) Smallpox still entombed? Lancet 1:1103

Oldstone MBA (1998) Smallpox. In: Viruses, plagues, and history. Oxford University Press, Oxford pp 27–44

Osler W (1876a) The initial rashes of smallpox. Can Med Surg J 5:241–255

Osler W (1876b) Haemorrhagic smallpox. Can Med Surg J 5:289–304

Osler W (1892) The principles and practice of medicine. Appleton, New York, pp 46–59

Pääbo S (1989) Ancient DNA extraction, characterization, molecular cloning, and enzymatic amplification. Proc Natl Acad Sci USA 86:1939–1943

Page DL (1953) Thucydides description of the plague. Class Q 3:97–119

Parry A (1969) The language of Thucydides' description of the plague. Bull Inst Class Stud 16:106–118

Procopius (1981) Histories of the wars. Dewing HB (translator), Loeb Classical Library, Harvard University Press, Cambridge

Raoult D, Aboudharam G, Crubézy E, Larrouy G, Ludes B, Drancourt M (2000) Molecular identification by "suicide PCR" of *Yersinia pestis* as the agent of medieval black death. Proc Natl Acad Sci USA 97:12800–12803

Reid AH, Fanning TG, Janczewski TA, Taubenberger JK (2000) Characterization of the 1918 "Spanish" influenza virus neuraminidase gene. Proc Natl Acad Sci USA 97:6785–6790

Ricketts TF (1908) The diagnosis of smallpox. Cassell, London

Roberts C, Manchester K (2005) The archaeology of disease, 3rd edn. Cornell University Press, Ithaca, NY

Rollo F, Luciani S, Canapa A, Marota I (2000) Analysis of bacterial DNA in skin and muscle of the Tyrolean iceman offers new insight into the mummification process. Am J Phys Anthropol 111:211–219

Shrewsbury JFD (1950) The plague of Athens. Bull Hist Med 24:1–25

Soupios MA (2004) Impact of the plague in Ancient Greece. Infect Dis Clin N Am 18:45–51

Spencer M, Howe CJ (2004) Authenticity of ancient DNA results: a statistical approach. Am J Hum Genet 75:240–250

Taubenberger JK, Reid AH, Krafft AE, Bjwaard KE, Fanning TG (1997) Initial genetic characterization of the 1918 "Spanish" influenza virus. Science 275:1793–1796

Thucydides (1919) History of the Peloponnesian War, books I and II. Harvard University Press, London, pp 341–359

Thucydides (1989) The Peloponnesian War. Cambridge University Press, New York

Tumpey TM, García-Sastre A, Taubenberger JK, Palese P, Swayne DE, Basler CF (2004) Pathogenicity and immunogenicity of influenza viruses with genes from the 1918 pandemic virus. Proc Natl Acad Sci USA 101:3166–3171

Zink AR, Reischi U, Wolt H, Nerlich AG (2002) Molecular analysis of ancient microbial infections. FEMS Microbiol Lett 213:141–147

Part I
The Techniques and Methods

Chapter 2
Identification and Interpretation of Historical Cemeteries Linked to Epidemics

Dominique Castex

Abstract Several types of event (wars, massacres, natural disasters, famines or epidemics) can lead to mortality crises resulting in the formation of funerary deposits unlike those found during more "ordinary" periods. This chapter specifically reviews the exploitation of demographic data from dental and bone remains to resolve the cause of a mortality crisis. Different age groups in a population are not affected in the same manner by all crises and it is therefore possible that the detection of possible anomalies in the demographic parameters among the archaeological series studied can be a useful indicator as to the origin of the deaths. This fact is illustrated by the analysis of three series in France in which palaeobiochemistry confirmed the presence of the Yersinia pestis plague bacillus. These results have allowed us to refine the methodological and analytical thematic study of both funerary archaeology and anthropology. Historical demographic analyses must be intensified in order to define more precisely the impact of different types of crisis on a population, thus deriving different typical profiles allowing interpretation of age and sex distributions and their possible anomalies. Analysis of osteological samples from periods of epidemic should cover as large a choice of sites as possible, both chronologically and geographically, in order to establish not only one "model" but several models illustrating crisis mortality.

2.1 Introduction

Certain events in the past (wars, massacres, natural disasters, famines or epidemics) have generated a great number of deaths and have led to veritable mortality crises. Although often studied historically, this theme, despite its rich potential, is relatively recent in the domains of archaeology and anthropology.

Dominique Castex
CNRS, UMR 5199 PACEA, Laboratory of Anthropology of Past Populations, Bordeaux 1 University, Avenue des Facultés, 33405 Talence cedex, France
E-mail: d.castex@anthropologie.u-bordeaux1.fr

D. Raoult and M. Drancourt (eds.), *Paleomicrobiology: Past Human Infections.*

In response to multiple deaths, the treatment of the bodies undertaken often results in the formation of funerary deposits unlike those found during more "ordinary" periods, e.g. several bodies in a single container, sometimes several structures juxtaposed. Once archaeological methods prove the simultaneity of the deposits (Duday 2005, 2006) and a phenomenon of abnormal mortality linked to a particular event is suspected, an interpretation can be attempted. The grave and, on a larger scale, the cemetery can become choice objects of analysis in understanding mortality crises of the past. On the one hand, tombs – valuable witnesses of cultural investment – can provide information about the reactions and specific treatments sometimes undertaken during periods of crisis while, on the other hand, skeletons represent a biological reality that can help clarify the nature of death.

The development of "preventive archaeology" during the 1980s, together with revised methods of approaching the excavation of burial deposits, have contributed to the discovery of several funerary deposits that followed epidemic crises from different periods. Some of these deposits have already undergone various analyses, which now allow interpretative hypotheses (Castex and Cartron 2007). Although important information linked to funerary archaeology contributes to the understanding of these particular burial deposits, the scope of this article is voluntarily limited to a specifically anthropological angle of analysis, specifically to the exploitation of demographic data from dental and bone remains. By using precise methodology and tools, the analysis of the composition of an archaeological population by parameters of age and sex is revealed as very pertinent in the interpretation of abrupt mortality crises due to epidemics.

2.2 Plague Cemeteries: From First Interpretations to the Identification of the Great Historic Plagues

2.2.1 *Analysis of Composition by Age and Sex: Some Methodological Reminders*

The first stage is the acquisition of the individual biological data, sex and age at death, of all the exhumed subjects[1]. These parameters are then used to define the composition of the population by age and sex as well as possible; a second stage must include the establishment of a mortality profile and the calculation of the rate of masculinity[2]. It is then possible to verify whether the distributions as a function

[1] These estimates must be as reliable as possible and on this subject the reader is referred particularly to an article by Bruzek et al. (2005). The methods applied to different deposits varied as they depended inevitably on methodological progress. Nevertheless, the series studied earliest have since undergone readjustments, which now allow reliable comparison. In addition, the representativeness of the subjects has been voluntarily limited to those of less than 30 years as the imperfection of age estimation methods for adults does not allow discussion beyond this threshold; however, more recent methods may be worth attempting later (Schmitt 2002).

[2] The rate of masculinity is the ratio of the number of men to the number of men and women; the theoretical rate is 50%.

of age and sex, obtained from the available archaeological samples, are close to those expected in the case of a natural demography[3] or, on the contrary, if they reveal anomalies connected to a specialisation that needs to be interpreted (Masset 1987; Sellier 1996; Blaizot and Castex 2005; Castex 2007).

In order to compare the data obtained to those expected in a situation of ordinary mortality[4] the different ages at death are distributed into 5-year groups (with the exception of the first two groups, of 1 and 4 years, respectively) of attained age in accordance with Ledermann's life tables (1969). To establish the subjects' mortality profile, a mortality quotient[5] is established for each age group and the quotients obtained are then compared to those of Ledermann (1969)[6].

The constitution by age and sex of several sites found in epidemic contexts will be analysed on the basis of methodological acquisitions fully developed elsewhere (Sellier 1996) and recently applied in a particular case (Castex 2005). The benefits of the analysis of age and sex parameters in palaeobiological studies needs no further demonstration and such analysis is of particular interest in the case of abrupt mortality crises. In fact, the different age groups of a population are not affected in the same manner by all crises, the nature of which will inevitably operate a selection in terms of age and sex, and it is therefore possible that the detection of possible anomalies in the demographic parameters among the archaeological series studied can be a useful indicator as to the origin of the deaths.

2.2.2 *Initial Analyses and Arguments in Favour of a Mortality Crisis Due to an Epidemic*

2.2.2.1 Saint-Pierre, Dreux, Eure-et-Loir (Fourteenth Century)

The excavation of Place Métezeau at Dreux within the framework of an urgent salvage operation by a team of archaeologists from the Centre Region (P. Dupont, in charge, and

[3] By this we mean a classical distribution of age and sex, i.e. as close as possible to that expected in traditional populations found in a schema of archaic or pre-Jennerian mortality before the Industrial Revolution (Masset 1975; Sellier 1996).

[4] As a reference of ordinary mortality, Ledermann's (1969) life tables were chosen.

[5] The mortality quotient is represented by aQx, where x is the age of entry into an age group and a is the time spent in years in that group. The mortality quotient concerns the number of deaths within an age group as a proportion of the numbers of a population likely to die within that age group initially, and thus represents the probability of death within a precise age group. It differs greatly from the rate of death, which shows the proportion of deaths within an average population, and is hence much more pertinent to the comparison of a theoretical natural mortality and a mortality obtained from an archaeological sample.

[6] Ledermann's tables (1969) allow the calculation of a confidence interval (at 95%) of the mortality quotients, as shown by a range of values on all the diagrams presented. I have chosen to present only those references related to a life expectancy at birth of 30 years, as this parameter lies between 20 and 40 years for known pre-Jennerian populations (Masset 1975; Sellier 1996). Ledermann's data are the essential element of discussion in any comparison of theoretical natural mortality and the demographic characteristics of our archaeological populations.

Fig. 2.1 A simultaneous burial at Dreux (Saint-Pierre) during the excavation. *Photograph* P. Dupont and U. Cabezuelo (Center Region Archaeological Service)

U. Cabezuelo) benefitted from the sporadic intervention of an anthropologist (P. Courtaud, UMR 5199). The first surprise was to discover graves completely atypical to those expected (Fig. 2.1). The continuing excavation revealed a very particular utilisation of the funerary space, with the presence of numerous multiple burials containing between 2 and 22 subjects, adults and immature individuals combined. The first date proposed for the utilisation of this part of the cemetery, based on stratigraphic data and on fragments of ceramics contained in the grave filling, was the twelfth century, but further dating by ^{14}C more precisely indicated the fourteenth century. The interest of this site lay in its unusual problems and in the possibility of exploiting this type of context from both an archaeological and an anthropological viewpoint for the first time. The presence of several simultaneous and contemporary burials suggested a mortality crisis, but which observations could be applied to justify this conclusion? Which anthropological tools would allow interpretation of this crisis?

In total, 22 graves yielding 72 individuals, of which 35 were adults and 37 immature subjects, were studied (Castex 1992, 1994, 1995; Cabezuelo and Castex 1994). The structure by age of those inhumed was studied so as to highlight possible differences from a natural demography (Fig. 2.2). The proportion of immature subjects within the total population is compatible with that of a theoretical mortality (51.4%). However, a detailed study of the non-adult age groups clearly indicates a "non-natural" population, i.e. a total absence of newborn infants and few individuals from the 1–4 year age group, contrasting with the growing mortality of the older

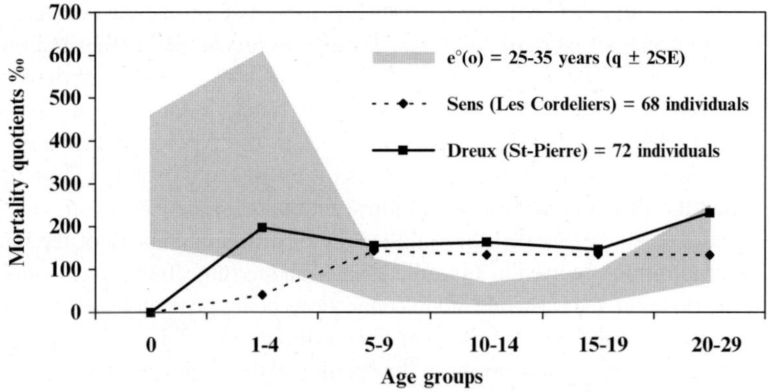

Fig. 2.2 Mortality profiles of immature individuals and young adults for Dreux (Saint-Pierre) and Sens (Le Clos des Cordeliers). Comparison with Ledermann's data (1969)

age groups from 5 to 19 years. The high mortality of the young adult age group (20–29 years) also differs sharply from that of a theoretical distribution obtained from typical tables (the proportion of young adults to that of the total adult population is 22.9%)[7]. In addition, the rate of masculinity of 72%[8] revealed a clear concentration of masculine subjects in this part of the cemetery. The particular nature of the graves encountered in the studied sector of the cemetery is thus associated with a specialised composition in terms of age and sex. The absence of precise stigmata on the skeletons allowed us to exclude violent death and, consequently, acts of war or massacre and oriented us towards the hypothesis of an epidemic[9]. The only noticeable pathological facts were, on the one hand, the abundance of dental tartar, which could be related to a particular type of alimentation implying a social connection between the individuals and, on the other hand, particularly frequent signs of anaemia, which seem to indicate that this human group suffered from numerous restrictions due to their surroundings (deficiencies, malnutrition, etc.) that could have presented an environment favourable to an epidemic.

2.2.2.2 Le Clos des Cordeliers, Sens, Yonne (Fifth–Sixth Century)

An archaeological salvage intervention carried out in 1989 by D. Maranski (the Sens municipal archaeologist) revealed structures for habitation as well as a funerary zone containing four multiple graves[10]. Unfortunately, the poor conditions of salvage did

[7] In a schema of archaic mortality this proportion evolves from 18 to 10% for life expectancies at birth when including individuals aged between 20 and 40 years, respectively.

[8] The difference from the theoretical distribution of 50% is significant at $P<0.05$.

[9] The diagnosis of an epidemic by analysis of bone remains is impossible as the rapid action of the infectious agents does not allow time for the development of osseous lesions, except in the case of those epidemic diseases that are non-lethal in the short-term, such as leprosy, tuberculosis and syphilis.

[10] This intervention completed those already undertaken in 1979 by J. Nicolle (archaeologist from Sens) and in 1985 by G. Depierre (TR Ministre de la Culture, UMR 5594).

not allow the recognition of the exact limits of these burials, which are also partly covered by elements of modern buildings. The graves are simple ditches hollowed in the earth without specific architectural elements apart from partial coverings of slabs from hypocausts in one case and large calcareous blocks in another. A first dating using stratigraphic arguments and with reference to the typology of an individual burial found in the same sector suggested the ninth–eleventh century. These burials have formed the basis of two research studies (Guignier 1996, 1997).

A taphonomic study revealed the simultaneity or near-simultaneity of the inhumations. Decomposition occurred in an infilled space, all the subjects appear in quite good anatomical condition, the so-called labile connections being generally well-preserved,[11] and entanglement of the bodies is seen at several levels (Fig. 2.3). These burials have now been radiocarbon dated to between the fourth and the sixth centuries.

The lowest number of individuals taken from these graves is 73, of which 45 were adults and 28 immature subjects. Although the proportion of immature subjects within the total population (38.4%) appears much lower than that of the site at Dreux; the distribution of age at death is, in many ways, relatively close (Fig. 2.2). Apart from the mortality quotient of the 1–4 year age group, which is clearly lower in the case of Sens, and that of the 20–29 year age group, compatible with a theoretical mortality[12], the similarity between the two death curves concerns the

Fig. 2.3 One of the four simultaneous burials at Sens (Le Clos des Cordeliers). *Photograph* D. Maransky (Sens Municipal Archaeological Service)

[11] The only movements observed are those directly linked to the synchronous decomposition of superimposed bodies (Duday 2005, p 198).

[12] The representativeness of adults under 30 years to that of the total adult population is 13.3%.

anomalies detected in the 5–9 year, the 10–14 year and the 15–19 year age groups. These groups are all over-represented and form an almost flat curve. The most noticeable difference concerns the rate of masculinity, which at Sens is 45.2%, very close to that of a natural demography. Supported by the total absence of pathological lesions, the hypothesis of a mortality crisis linked to an epidemic of an unknown nature, as in the case of Dreux, seems quite plausible.

2.2.2.3 Additional Arguments for Mortality Crises by Epidemic

The archaeo-anthropological funerary data and stratigraphical data, at both Dreux and Sens, allow us to interpret the adaptation of a community to a phenomenon of abrupt mortality. It is interesting to note that, in both cases, the laying down of the bodies reveals a relatively well-ordered administration and consequently gives a different picture to that of the disordered burial ditch one might associate with the context of a mortality crisis[13].

In addition to archaeological data, biological data, rarely exploited until now in such contexts, has provided arguments in favour of the hypothesis of multiple burials probably due to mortality crises linked to epidemics of an unknown nature. In spite of the absence of any precise historical records, the tools of biological anthropology have proved their utility in the interpretative process, in particular by detecting demographic anomalies, a non-negligible argument in any discussion on the origins of death.

For such little- or non-documented periods, it is important to insist on the fundamental necessity of accurate dating. The site at Sens is exemplary because the definite dating acquired during the analysis was of fundamental importance in that it allowed a reorientation of historical research and, consequently, raised the possibility that there was a relationship between the multiple burials of Le Clos des Cordeliers and the epidemic of plague that affected the town in 571 A.D. (Guignier 1997).

2.2.3 New Aspects from Archaeological and Historical Sources

2.2.3.1 A Case of Affirmed Plague: Les Fédons, Lambesc, Bouches-du-Rhône (Sixteenth Century)

An archaeological operation on the site of Les Fédons at Lambesc (north-west of Marseille) revealed 101 inhumed tombs. The excavation of this site, undertaken by a team of archaeologists from AFAN[14] under the direction of P. Reynaud, covered the entire cemetery. Burials were of two types: individual, of which there were 75,

[13] Cf. the recent study by Ph. Blanchard (2006).

[14] Association pour les Fouilles Archéologiques Nationales, now INRAP (Institut National de la Recherche Archéologique Préventive), since February 2002.

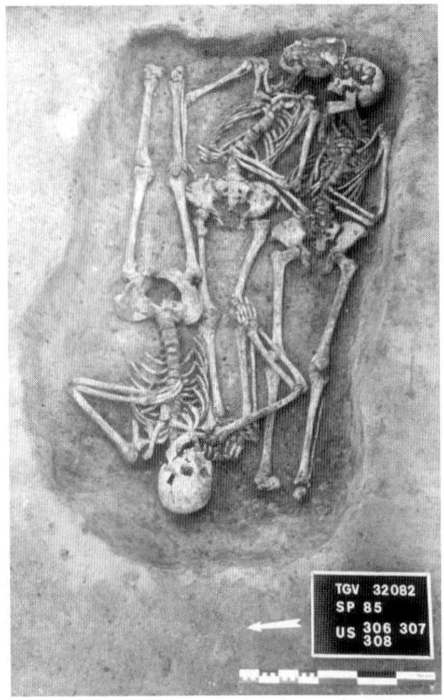

Fig. 2.4 A triple burial at Les Fédons (Lambesc, Bouches-du-Rhône). *Photograph* P. Reynaud (INRAP)

and multiple, of which 21 were double, 4 triple (Fig. 2.4) and 1 quadruple. As in the initial analysis of the site at Sens (Guignier 1996), the deposits at Les Fédons formed the basis of an evaluation and an archaeological report (Bouttevin et al. 1996; Reynaud et al. 1996) and has recently benefitted from an exhaustive publication (Bizot et al. 2005). Osteological observations realised in the field revealed that the deposits inside those graves containing several individuals were simultaneous, and archaeological data indicates an orderly and intelligent management of the cadavers (Moreau et al. 2005; Reynaud and Bizot 2005).

The great originality of this site compared to other known funerary contexts in times of epidemic, is that of the presence of individual graves alternating with double, triple, and one quadruple graves. This shows that a mortality crisis may generate individual graves – a point that must be taken into account in the global study of a site.

From the beginning, in order to optimise the biological data in the field, an effort was made to systematically consolidate bone remains, particularly those of the coxae for sex estimation. Of the 133 individuals inhumed, 72 immature subjects and 61 adults, 32 women and 29 men were counted. The profile of distribution of age at death, obtained from calculation of the mortality quotients, revealed numerous apparent anomalies compared to a theoretical mortality (Castex 2005) (Fig. 2.5). The infantile mortality quotients, which concern the 0–1 year and the 1–4 year age groups are very low compared to those found in the framework of a natural mortality,

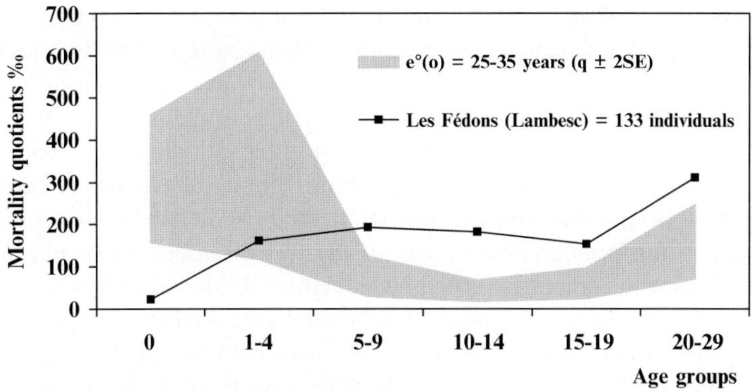

Fig. 2.5 Mortality profiles of immature individuals and young adults for Les Fédons (Lambesc, Bouches-du-Rhône). Comparison with Ledermann's data (1969)

showing an imbalance for children under 5 years – a deficit particularly noticeable for those under 1 year. On the contrary, calculations of the mortality quotients for the age groups 5–9 years, 10–14 years and 15–19 years show a clear over-representation of these groups, without the expected relationships, e.g. a minimum for the 10–14 year age group[15]. Another anomaly is the over-representation of young adults (20–29 years) within the total adult population (i.e. 31.3%); analysis of death distribution as a function of sex shows that this over-representation is specifically linked to a high female mortality within this age group. However, the sex distribution of adults as a whole remains equivalent to that of a natural demography, with the number of male subjects being equal to that of female subjects; the raw figures give a rate of masculinity of 47.5%, which conforms to the statistically theoretical rate of 50%. Thus, apart from the proportion of immature individuals within the total population (54.1%) and the sex distribution, both of which being compatible with that of a theoretical distribution, the age at death distribution of those under 30 years shows a clear distortion compared to that of a natural demography. In order to investigate whether the peculiarities observed at the site of Les Fédons could be considered characteristic of an epidemic of plague, we consulted several historical demographic studies of times of plague.

2.2.3.2 Exploitation of Archival Sources and Other Archaeological Data

The principal documents that can by utilised in relation to the estimation of mortality rates by age are those provided by the work of Hollingsworth and Hollingsworth (1971), concerning the London plague of 1603, and that of Mallet (1835), concerning plague in Geneva throughout the seventeenth century; these studies have also been

[15] In the case of a normal mortality, the age group for which the number of deaths is the lowest.

used by Biraben (1975). Using the raw numbers of deaths provided by the authors, mortality quotients for each age group[16] were calculated (Fig. 2.6). In both London and Geneva, the distribution of mortality quotients by age group during periods of plague is completely disturbed compared to that of a normal mortality: the infantile mortality quotient is very low and, in contrast, there is a clear super-mortality of young children, adolescents and adults during periods of plague. A comparison with the cholera epidemic that struck Paris in 1832 (Mallet 1835) shows important differences in the demographic impact of the two diseases: unlike plague, additional deaths due to cholera are little noticed during childhood and adolescence and become progressively more and more important during adulthood, especially in the oldest groups. These results are confirmed elsewhere by other analyses (Faron 1997). These studies provide similar profiles, always with the same variations between a classical mortality and mortality during periods of plague, a profile comparable to that obtained for the osteological series of Les Fédons. In fact, the profile of mortality by plague appears very close to the profile of a living population, thus demonstrating the non-selection in terms of age of the victims of *Yersinia pestis* (Castex 1996, 2005). Other historical demographic studies covering plague in Provence in the eighteenth century (Signoli et al. 2002; Signoli 2005) also show a clear difference between the demographic profile of ordinary mortality and that of plague.

We also considered other archaeological series from known epidemic contexts and, more precisely, two burial ditches discovered in 1994 of victims of the great

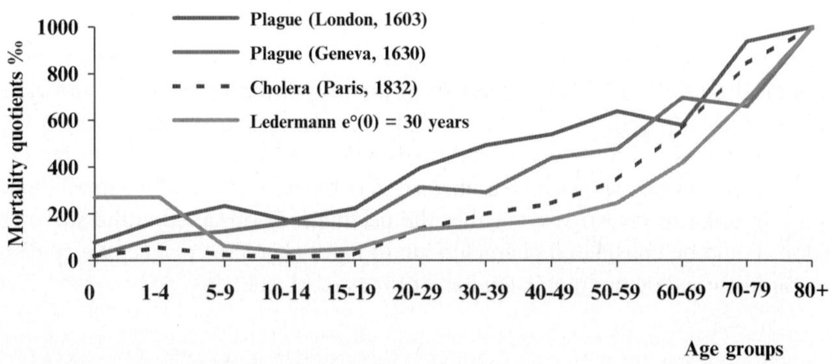

Fig. 2.6 Comparison of probabilities of death for different epidemics (data from Biraben 1975; Hollingsworth 1971; Mallet 1835) and comparison with Ledermann's data (1969). *Graphical representation* P. Sellier (UMR 5199) and D. Castex

[16] The data from the registers allowed an examination of the distribution of adults. These were sorted into decennial groups, which is difficult to achieve for adults over 30 years in the case of studies of archaeological series.

epidemic of plague of 1720–1722: the ditches L'Observance at Marseille and Le Délos at Martigues (Signoli 1998). Although the methodology and tools differed from ours, the data obtained from these two sites, particularly Martigues, revealed many points in common with observations made at Les Fédons. At L'Observance, the imbalances observed were distinctly less important and, apart from the deficit in the youngest age group, there was not the large proportion of children and older adolescents seen at Les Fédons. In an attempt to explain the differences between the mortality profile of Marseille and that of Martigues, an epidemiological hypothesis (Signoli 1998, 2005) has been proposed: the site of L'Observance may have been established during the recurrence of the epidemic in 1722[17], while that of LeDélos was established at the peak of the epidemic. Other interesting comparisons, based on archaeological series, between ordinary deaths and deaths linked to plague show demographic peculiarities, certainly linked to the epidemic impact, but also inherent to the constitution of the archaeological samples (Margerison and Knüsel 2002).

The few available historical and archaeological documents reveal comparable mortality profiles by plague as a function of age; however, as a function of sex, the results seem much more contradictory. These differences, which are difficult to understand from a medical point of view, could be linked to exposure to the disease (Biraben 1975). The high female mortality among young adults seen at Les Fédons could thus be due to the large number of young women employed at the infirmary of Les Fédons, as revealed in the archives (Rigaud 2005). Some historical studies corroborate this excess of females (Signoli 2005), whereas others tend to show an excess of males within the adult population, as in the parishes of Paris during the fourteenth century plague (Lucenet 1985), and in London during the plagues of 1603 and 1625 (Biraben 1975).

Following these archaeological discoveries, investigations within the scope of molecular palaeobiochemistry were rapidly undertaken. Residues of dental pulp from the L'Observance (Marseille) and Les Fédons (Lambesc) sites were able to provide ancient DNA sequences of the bacillus *Yersinia pestis*, the plague vector (Drancourt et al. 1998, 2005).

Within the fields of archaeology and biological anthropology, the site at Les Fédons thus appears a quite original funerary example; this site benefitted from an exhaustive excavation[18] and the inhumed population is perfectly dated, with a representative number of inhumations where the cause of death is known (archival sources and molecular palaeobiochemistry). The analysis of this site, in conjunction with available references to historical data, as well as comparison of the results with those obtained in other studies of cemeteries linked to plague, have allowed a very precise clarification of this mortality crisis.

---

[17] Hypothesis founded upon a comparison of the demographic characteristics of the exhumed osteological sample and data from the records of the convent at L'Observance for 1722.

[18] This is particularly important as, in many cases, a non-exhaustive excavation can be held responsible, at least in part, for a sketchy interpretation of certain demographic anomalies.

2.2.4 *Identification of the Black Death and Justinian Plague*

The results obtained on the demographic impact of plague inevitably led to questions about previously studied funerary sites, e.g. Saint-Pierre at Dreux and Le Clos des Cordeliers at Sens (see above), which present an abnormal mortality and for which historical sources are lacking. The mortality quotients obtained at Dreux and Sens, when compared with those of Les Fédons, reveal striking similarities between the three sites (Figs. 2.2, 2.5): the anomalies registered between a mortality by plague and a natural mortality were very close to those observed between the two ancient series and Ledermann's theoretical data (Castex and Friess 1998; Sellier and Castex 2001). Although initially unable to affirm the impact of plague on the basis of this analysis alone, we nevertheless considered it a non-negligible argument, but requiring corroboration by other analyses for a final diagnosis.

Further molecular palaeobiochemistry research was undertaken (Drancourt et al. 2004) and the presence of the *Yersinia pestis* bacillus was confirmed in the two archaeological series: the Black Death at Dreux – thus corroborating the results obtained at Montpellier (Raoult et al. 2000) – and the "Justinian" plague at Sens, revealing for the first time the presence of the plague bacillus in the sixth century (Castex and Drancourt 2005). These results allowed the clarification of a major historical problem as the third pandemic of plague, which began in South-East Asia at the end of the nineteenth century, was the only one to have had a sure microbial origin clearly identified by Yersin in 1894. In particular, the identification of *Yersinia pestis* as being responsible for the Black Death ends the controversy as to its etiology, ruling out other pathogens that had previously been incriminated (Scott and Duncan 2001).

The following illustrates the importance of the results obtained: sites where the historical records attest an episode of plague allow an epidemic model or models to be confirmed, in turn allowing hypotheses of a particular epidemic crisis to be proposed at other sites where historical data is unavailable. Archaeological studies are thus orientated towards further indications and towards the search for further data (e.g. re-evaluation of the date of a site considered as known, molecular palaeobiochemistry research, etc.). These results and the distance necessary for their objective interpretation have allowed us to develop more precise methodological and analytical thematic studies in both funerary archaeology and anthropology, leading to a more systematic approach to future discoveries and studies and the co-ordination of interdisciplinary collaboration that is fundamental to the understanding of such contexts[19].

[19] This problem was developed within the framework of the quadrennial project 2003–2006 of La Maison des Sciences de l'Homme d'Aquitaine, specifically that on mortality crises. Several new sites linked to epidemics have been studied. Others are undergoing analysis.

2.3 "Possible" Plague Cemeteries: Epidemic Impact and/or Initial Selection

2.3.1 Saint-Benedict of Prague (Late Sixteenth Century): a Previously Selected Population?

The vast cemetery of Saint-Benedict of Prague, excavated in 1971, contains more than 800 graves, of which many are multiple inhumations[20]. Most of the skeletons from these burials were the subject of an anthropological study in 1988 (Hanakova and Stloukal 1988)[21]. The archaeological level on which we concentrated concerns the latest phase of the cemetery. This phase corresponds to a large number of inhumations (about 450, i.e. more than one-half of the total number of inhumations) which, according to initial examination of the records, could be linked to the plague of 1680, at which time the cemetery and structural elements discovered on the site belonged to the Premonstratensian order. By its very nature, the cemetery of Saint-Benedict of Prague was likely to introduce further elements warranting reflection into the particular context of acute mortality crises linked to plague (Castex et al. 2003, 2005). As the number of individuals was very large we concentrated initially on an exclusive study of the multiple graves[22].

To date, 20 multiple graves, containing 120 subjects, have been studied (Fig. 2.7)[23]. Compared with data from Ledermann's typical tables (1969), the mortality quotient curve shows various flagrant anomalies (Fig. 2.8). Firstly, there is a very clear under-representation of children under 5 years (and even under 10 years), with a total absence of the former age group. In fact, the number of immature subjects, as a proportion of the total population, is low at only 27.4%. The age groups 5–9, 10–14 and 15–19 years show a clear disproportion between each other, with a slow inflation of the mortality quotients. Finally, we see a peak of mortality in the 15–19 year and, above all, in the 20–29 year age groups, the representation of young adults within the total adult population being 47.8%. The rate of masculinity is particularly high at 83.6%.

It would therefore seem that the criteria of age and sex show a very selective composition: the mortality quotient curve, very different to that observed in the

[20] A large amount of archaeological data was made available. This data resulted from the report produced by B. Martinec, a Czech archaeologist in charge of the excavation (Martinec 1971).

[21] The aim of this study was to obtain general biological information on the population of Prague from the Middle Ages until a recent period and no distinction was made between those inhumed in individual graves and those in multiple graves.

[22] The long-term objective is to study all the graves in order to compare, for the same chronological period, mortality as a function of age and sex between two types of 'recruitment': one more or less 'natural' and the other linked to a mortality crisis.

[23] We have chosen to present a triple grave, which, with double graves, occurs most frequently, although at least three graves contain the remains of 9 subjects and another contains up to 20 (the deposits were organised in successive layers in very deep and narrow ditches.)

Fig. 2.7 A simultaneous burial at Saint-Benedict (Prague, Czech Republic). *Photograph* V. Martinec (Prague Archaeological Service)

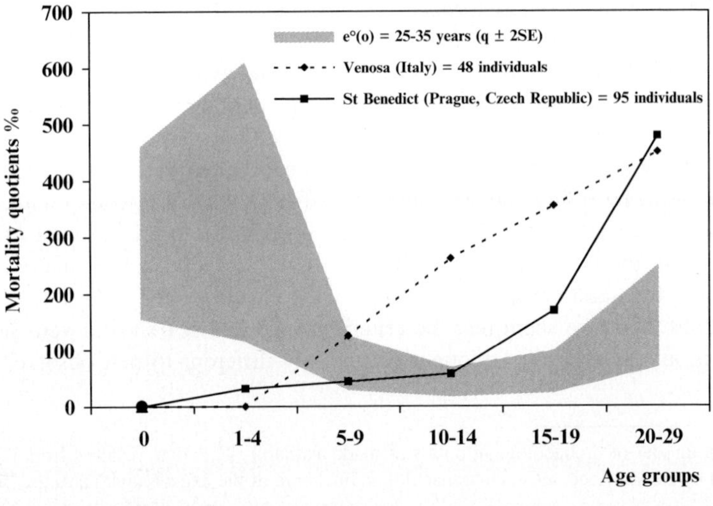

Fig. 2.8 Mortality profiles of immature individuals and young adults for Saint-Benedict (Prague, Czech Republic) and Venosa (Lucania, Italy). Comparison with Ledermann's data (1969)

case of a natural mortality, also differs from that expected in the case of plague, especially in the relationship between the quotients of the 5–9, 10–14 and 15–19 year age groups and in the excessive numbers of individuals aged 20–29 years. This divergence in the profiles invites various comments. The hypothesis of an epidemic must be retained as it is supported by the archaeological facts, the suddenness of the deaths having caused the establishment of multiple graves, and the absence of specific lesions on the skeletons. Even if we are dealing with an epidemic of plague, this fact alone would seem unable to explain such large anomalies in the distribution of age and sex at death. In addition to the epidemic factor and its virulence, other mechanisms that may have contributed to this abnormal distribution of deaths must be invoked. An explanation must be found in the constitution of the original group. This requires the use of historical records, which alone are able to specify the status of the site and allow recognition of the existence of a possible relationship between the sector of multiple graves and the Premonstratensian monastery (a group of individuals selected according to sex with a large majority of young men?). Perhaps a different type of epidemic occurred, imposing once again the need to access textual sources in which a precise incident – maybe less noteworthy than plague, but nevertheless recorded in written form – may be identified.

2.3.2 *Venosa, Lucania, Southern Italy (Eighth–Tenth Centuries): the Nature of the Crisis Reconsidered*

At the site of Venosa, excavated in 1986 and 1987, five adjacent graves, dated to between the eighth and tenth centuries, each containing between 7 and 12 individuals (a total of 48 subjects) placed side-by-side or one above the other were discovered. The presence of several simultaneous inhumations led to the hypothesis of a mortality crisis (Fig. 2.9). Although no documentary source mentions it, the possibility of an

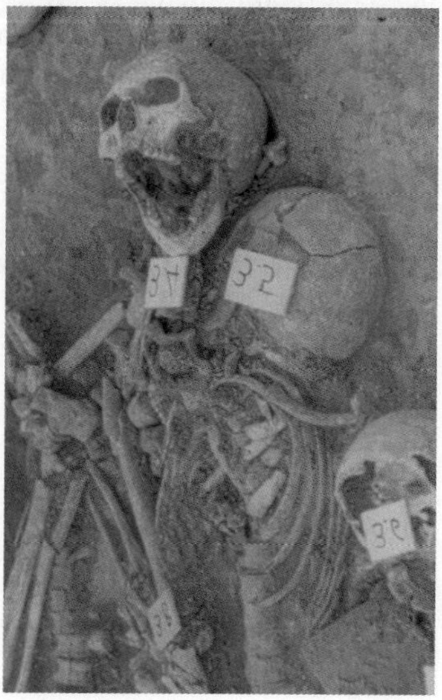

Fig. 2.9 A simultaneous burial at Venosa (Lucania, Italy). *Photograph* R. Macchiarelli, L. Bondioli (Pigorini Museum, Rome)

epidemic of plague was discussed because of the absence of specific lesions on the skeletons and because of the structure by age and sex of the osteological sample compared with those of models elaborated in historic demography. The deaths of the subjects had previously been distributed into 10-year age groups (except for the first group, 0–4 years, of which there were none). The over-representation of the groups 5–14 and 15–24 years registered for the burials at Venosa showed a mortality profile very different to that of a natural demography but comparable, in general terms, to that observed in periods of plague (Macchiarelli and Salvadei 1989). The latter point is the only argument supporting the interpretation of a possible occurrence of plague in this case, as the other two arguments would equally apply to any epidemic that causes a large number of deaths and that acts so rapidly that osseous lesions do not occur. Because of the small size of the sample, it is important to have as precise an age estimation as possible for immature individuals and their distribution into age groups, so as to reveal true demographic anomalies – the only element that can inform us about the nature of the crisis that affected these individuals. The mortality profile obtained from the authors' raw data revealed several anomalies, which finally appear quite different to those observed in known cases of plague (Fig. 2.8). The proportion of immature subjects within the total population is 58.3%; although large, this value remains compatible with that of a theoretical mortality for a life expectancy at birth of 30 years. The 0–1 and 1–4 year age groups are totally imbalanced with a complete absence of subjects. The older age groups show a very regularly growing curve, with abnormally high numbers for those of 10–14, 15–19 and 20–29 years. A very clear over-representation of young adults as a proportion of the total adult population (45%) was noted. Thus, when studied more precisely, the ratios between the different immature age groups is very different from that generally observed in confirmed cases of plague: should this mortality profile be considered representative of plague? In this case should the anomalies observed be attributed to the existence of a population already selected by age (perhaps a selection in part of the site only)? Or should, as for Saint-Benedict of Prague, the validity of the first diagnosis be questioned and an epidemic crisis of a different nature be envisaged?

2.3.3 Further Lines of Research

The analysis of age and sex distributions in the sites of Saint-Benedict and Venosa has thus revealed both quantitative and qualitative details that differ from those generally identified in the context of plague. This new data invites us not only to re-examine historical hypotheses, perhaps accepted too quickly, but also to take into account the existence of human behaviour, too often simply ignored yet capable of introducing numerous imbalances into the consideration of archaeological populations.

It is therefore necessary to undertake additional studies of the two sites. A re-examination of the dates already proposed is required as well as a greater use of historical sources, especially at St. Benedict, where much more information ought to be available. In both cases, the interest of molecular palaeobiochemistry analyses becomes evident, e.g. the possibility of finding a pathogen different from that of plague[24] and thus proving, perhaps, that plague in the past was not necessarily linked to the action of *Yersinia pestis*. Within the framework of research on the validity of plague diagnosis, two more funerary sites, already considered promising in the long-term, may enrich the corpus available: the multiple burial at Gerasa, Jordan, may be linked to a seventh century plague (Seigne 2007), and the cemetery of the Santa Clara convent at Palma de Majorque, Balearic Islands, implicated by historical sources in the Black Death of the fourteenth century.

2.4 Other Cases of Cemeteries Linked to Mortality Crises Due to Epidemic

2.4.1 Issoudun, Indre (Seventeenth–Eighteenth Centuries): Epidemic Coupled with Famine?

In the context of a "preventive" excavation undertaken by INRAP from May to September 2002, 14 multiple graves were discovered in part of the ancient cemetery of Issoudun (Indre). These graves, dating from the late seventeenth to early eighteenth centuries, are grouped in a zone particularly dense with skeletons, resulting from the intensive use of a funerary space that functioned over a long period (thirteenth–eighteenth centuries). The funerary topography shows that the graves are aligned in relatively clear rows, all except two in the same orientation. The peculiarity of the context and the possibility of direct intervention in the field from the start of the excavation[25] (as at the site of Les Fédons, Lambesc, see above) allowed the use of recording methods adapted to a salvage excavation while favouring the maximum yield of information available to a post-excavation study, particularly of this type of site (Blanchard et al. 2003a, 2003b).

Apart from one double burial, the graves contained between 13 and 22 individuals, deposed simultaneously, and were composed of adults of both sexes and immature subjects showing a remarkable proportion of children over 1 year (Fig. 2.10). All the graves revealed a very rational organisation of the deposits according to age criteria (Fig. 2.11). On the basis of the arguments described above, the hypothesis of an epidemic was quickly formed, although its nature could not immediately be proposed. The minimum number of individuals was estimated at 203. The number

[24] Work has been engaged in this direction with Michel Drancourt, Unité des Rickettsies, CNRS UMR 6020, Faculté de Médecine, Marseille.

[25] Thanks to I. Souquet-Leroy, anthropologist in the field in charge of the excavation and study of the graves.

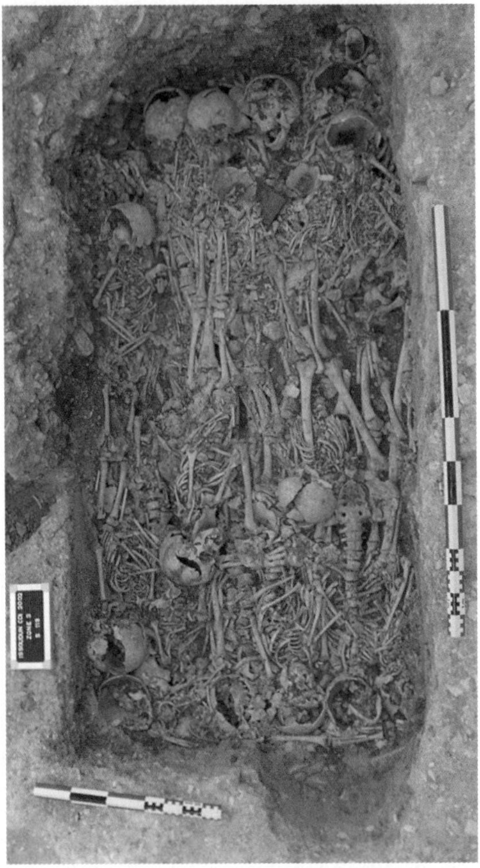

Fig. 2.10 One of the 14 simultaneous burials (S. 119) at Issoudun (Indre) containing 22 individuals. *Photograph* F. Porcell (INRAP)

of the archaeological sample, i.e. victims of this epidemic, was established from the total number of individuals from the multiple graves, both complete and incomplete, but also by taking into account some single graves suspected of being contemporaneous with the multiple graves[26]. The mortality curve revealed very noticeable anomalies compared with that of a natural population (Fig. 2.12). The numbers of 0- to 1-year-olds are very low whereas the numbers of 1- to 4-year-olds are high, although compatible with Ledermann's highest theoretical values (1969). The most surprising imbalance is the high number in the 5–9 year age group, followed by that of the 10–14 year age group. The corresponding quotients also show an over-representation. The proportion of immature subjects within the total population appears extremely high (76.4%).

[26] Their alignment with the multiple graves and the unusual deposit of the bodies inside the graves, e.g. several individuals lying on their stomachs, tended to prove their creation at a time of crisis.

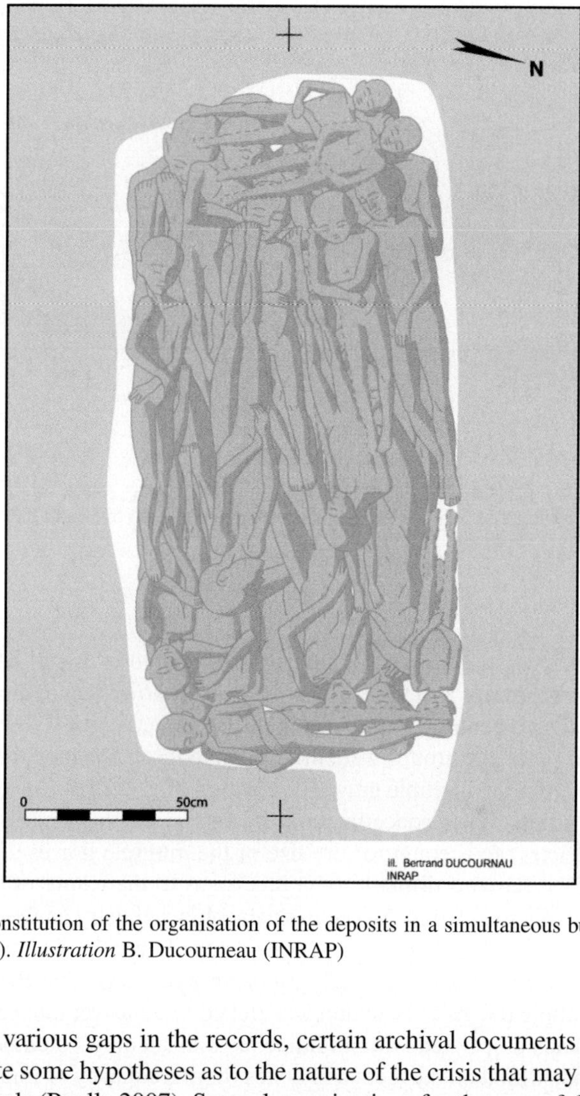

Fig. 2.11 Reconstitution of the organisation of the deposits in a simultaneous burial (S. 119) at Issoudun (Indre). *Illustration* B. Ducourneau (INRAP)

In spite of various gaps in the records, certain archival documents have allowed us to formulate some hypotheses as to the nature of the crisis that may have affected these individuals (Poulle 2007). Several examinations for the years following 1650, when excess mortality was particularly severe (war, famine and numerous diseases), were undertaken. For each of these years we were able to calculate different demographic parameters (proportion of immatures within the total population, rate of masculinity, etc.). Several comparisons were then made between crisis and non-crisis years and the findings were set against the results obtained from Issoudun's archaeological sample. Initially, two periods of crisis were targeted[27]: the years

[27] They had been skilfully analysed by regrouping the deaths into age groups comparable to those of our archaeological sample.

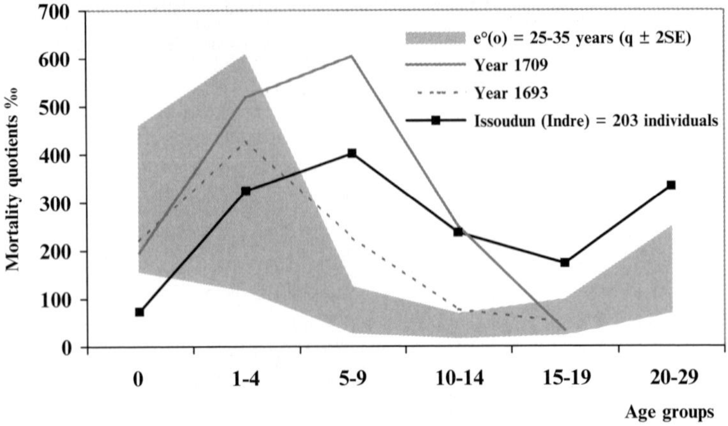

Fig. 2.12 Mortality profiles of immature individuals and young adults for Issoudun (Indre). Comparison with Ledermann's data (1969) and historical records for 1693 and 1709. *Documentation* P. Poulle (INRAP)

1693–1694 (various crises linked to the end of the reign of Louis XIV) and 1709 (well-known to historians as a year of poor harvests followed by famine and numerous diseases). Our attention was finally held by the crisis of 1709 as the ratios between the different age groups were the closest to those obtained for the archaeological sample from the multiple graves discovered (Fig. 2.12)[28]. Another argument for the choice of this crisis concerns the daily rate of death found in the archives, which was sufficient to account for the size of the multiple graves discovered[29]. If this is the case, it remains difficult to be precise as to the nature of the crisis that affected Issoudun's population. Molecular palaeobiochemistry analyses are, for the moment, negative[30] for the first pathogens researched (smallpox, plague, measles). In spite of this we maintain the highly probable hypothesis that the crisis which caused the multiple deaths at Issoudun was linked to an as yet unidentified human pathogen, perhaps associated with a famine, as indicated by a number of signs of growth stress identified on the teeth and bones (high frequency of linear hypoplasia of the dental enamel and numerous cases of cranial hyperostosis and rickets).

[28] When using the registers, no distinction could be made between young and older adults, which of course excluded a comparison of the 20–29 year age group.

[29] For example, in September 1709 there were up to 22 deaths on the same day.

[30] Report on palaeomicrobiologic analyses by L.V. Dang and M. Drancourt (UMR 6020, Faculté de Médecine, Marseille) within the framework of the Final Document of Synthesis on the site of Issoudun, at present being finalised. The absence of the pathogens researched does not, however, exclude the possibility of their having existed.

2.4.2 Boulogne-sur-Mer, Pas-de-Calais (Eighteenth Century): Hypothesis of a Smallpox Epidemic?

Seven multiple graves were found at a "preventive" excavation at L'Ilot Saint-Louis, Boulogne-sur-Mer, in all a total of 39 individuals. These graves were dated early eighteenth century (Belot and Canut 1995) (Fig. 2.13). The simultaneity of the osseous deposits associated with the contemporaneity of the different structures as well as the recurrence of the phenomenon led us to interpret this site as the result of an abrupt mortality crisis (Réveillas 2005; Castex and Réveillas 2007). It was possible to eliminate the hypotheses of war and famine in favour of that of an epidemic on the basis of historical, archaeological and anthropological arguments. The study of the ratios of the different immature age groups is worthy of attention. Except for a few details, the mortality profile overall appears close to that of a "natural" demography (Fig. 2.14). The number of adults is almost identical to that of immature subjects but the mortality quotient of children under 1 year is very low, whereas the age groups 15–19 years and 20–29 years show high numbers compared with those observed in a theoretical population. According to Ledermann's data (1969), the relation expected between the age groups 5–9, 10–14 and 15–19 is

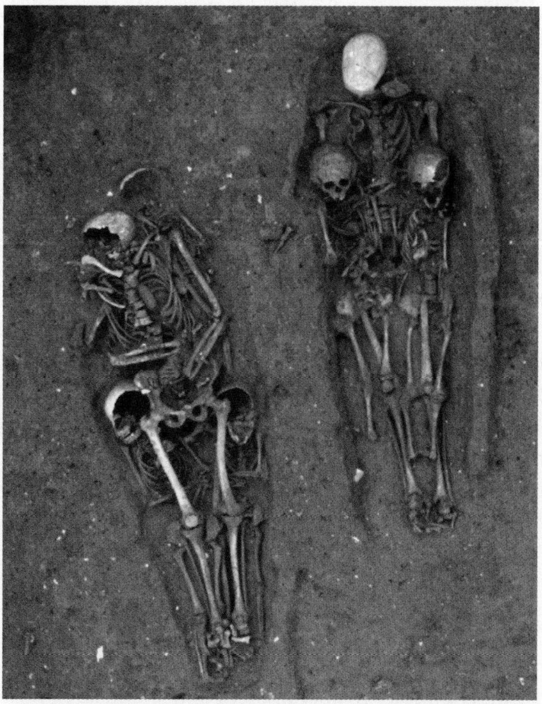

Fig. 2.13 General view of two simultaneous burials at Boulogne (Pas-de-Calais). *Photograph* E. Belot (Boulogne-sur-Mer Municipal Archaeological Service)

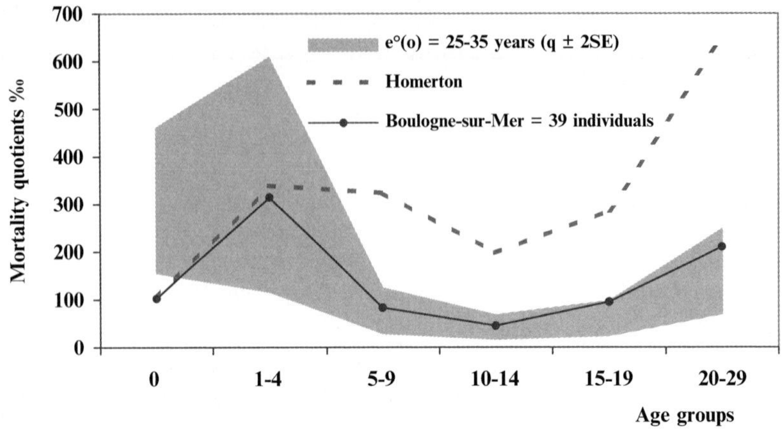

Fig. 2.14 Mortality profile of Homerton's hospital population and the archaeological population of Boulogne-sur-Mer. Comparison with Ledermann's data (1969)

respected, a minimum being found in the 10–14 year age group. The proportion of young adults is high, representing 21.05% of the total adult population. At 52.6%, the rate of masculinity conforms to that of a natural mortality.

These specific features, considered in regard to various historical, medical and demographic sources (Darmon 1986; Leca 1982), allow the elimination of certain epidemics, in particular plague. The data have been compared to that of a hospital at Homerton in the United Kingdom specialising in the treatment of smallpox (Razell 1977) (Fig. 2.14). Although the mortality quotients from Homerton are noticeably higher than those at Boulogne, the similitudes registered (low mortality quotient for children under 1 year and contrastingly high for the 1–4 year age group; relationships between the next three older age groups) perhaps indicate that smallpox is an interesting research path to follow, although other diseases, poorly documented but recurrent at that time (influenza, prickly heat, malaria), cannot be totally eliminated. Further research must be undertaken before proposing a definitive interpretation, in particular a more complete study of Boulogne's archives and additional inquiries into the demographic impact of smallpox and other types of epidemic, but also by molecular palaeobiochemistry analyses allowing the identification of specific micro-organisms.

2.5 Conclusions

In the domain of palaeobiology the parameters of age and sex of osteological series, far from being considered tools for the demographic reconstruction of past populations, appear more to furnish details, analysis of which is essential to the interpretation of the functioning of these sites. In the particular case of mortality crises, it is possible to demonstrate global models that can be appropriately applied to certain crises. Thus, even in the absence of historical data, a hypothesis of a particular demographic crisis can be proposed.

Using the example of plague, we have followed the thematic strategy of palaeobiology, the ultimate aim of which is a better comprehension of past mortality crises, from its debut with the intrinsic characteristics of a site. This field of study has obviously been encouraged by the efforts made on more recent sites, which benefit from the availability of historical sources. However, apart from exceptional cases, such as the epidemic of plague at Marseille in the early eighteenth century where both an osteological collection, outcome of an affirmed plague, and historical demographic data relative to this epidemic (see above) are available, to try to make archaeological facts and historical events coincide systematically is somewhat questionable and the results obtained can in no way represent a "model" that could be applied unconditionally to more ancient osteological series whose initial constitution is unknown. For such series, the only possibility of identifying anomalies in comparison with as large a natural distribution as possible is to refer to the theoretical models proposed by standard tables, in this case those of Ledermann (1969), although other tables exist.

Although by now we have accumulated a great deal of experience with plague epidemics, we have seen that mechanisms other than that of the epidemic and its virulence can lead to an abnormal distribution of deaths and sex. As well as the indispensable analysis of archaeological facts, this latter factor requires that the constitution of the original population be discussed, even if this means, in some cases, calling into question the nature of the crisis proposed by historical sources. Types of crisis other than plague epidemics can be considered, as we have seen with some of the examples discussed above that must remain at the hypothesis stage.

In order to progress, this subject needs the continuing and indispensable to-and-fro between the recently available methodological expertise and the ancient (osteological) series, and the two complementary approaches must be conducted in parallel. Firstly, historical demographic analyses must be intensified so as to approach more exactly the impact of different types of crisis and thus allow different typical profiles to be proposed with which to interpret age and sex distributions and their possible anomalies. Secondly, the analysis of osteological samples from periods of epidemic must depend upon as large a choice of sites as possible, both chronologically and geographically, in order to establish not just one "model" but several models illustrating crisis mortality.

Acknowledgements Other than those mentioned in the text who have all contributed, in one way or another, to the elaboration of the archaeological data indispensable to the palaeobiological studies that followed, my thanks go particularly to Patrice Courtaud (IR, UMR 5199), Germaine Depierre (TR, Ministère de la Culture, UMR 5594) and Bruno Bizot (Conservateur du Patrimoine, SRA PACA) for the sites of Dreux, Sens and Les Fédons, respectively. Through their knowledge of the field and of archaeo-anthropological problematics, they have given me the opportunity, from the early 1990s in the case of Dreux, to commit myself to the study of mortality crises. Thanks also to Frances Holden for her translation.

References

Belot E, Canut V (1995) Boulogne-sur-Mer. Ilot Saint-Louis. Opération archéologique préalable à l'aménagement du parking de l'Université du Littoral. Bilan des recherches, Service Archéologique Municipal de Boulogne-sur-Mer, Boulogne-sur-Mer

Biraben JN (1975) Les hommes et la peste en France et dans les pays européens et méditerranéens. EHESS, Centre de Recherches Historiques, vols 1–2, Mouton, Paris

Bizot B, Castex D, Reynaud P, Signoli M (2005) La saison d'une peste (Avril–Septembre 1590). Le cimetière des Fédons à Lambesc. CNRS, Paris

Blaizot F, Castex D (2005) Du bon usage des outils anthropologiques à l'étude des sociétés historiques. In: Dutour O, Hublin JJ, Vandermeersch B (eds) Origine et evolution humaine. Comité des Travaux Historiques et Scientifiques, Paris, pp 259–279

Blanchard Ph (2006) La gestion des cadavres en temps de crise. Exemples archéologiques médiévaux et modernes en Europe de l'Ouest. Masters Thesis, Université François-Rabelais de Tours

Blanchard Ph, Castex D, Souquet-Leroy I, Poulle P (2003a) Issoudun, centre de l'image et du multimédia. Archéopages 9:40–41

Blanchard Ph, Castex D, Souquet-Leroy I, Poulle P (2003b) Les sépultures multiples d'Issoudun (Indre, XVIIe s.): stratégie d'intervention et objectifs de recherche. Réunion scientifique de la Société d'Anthropologie de Paris, 1828ème séance, Paris, 16–18 January 2003, pp 199–200

Bouttevin C, Castex D, Moreau N, Reynaud P, Rigaud P (1996) Le cimetière moderne d'une infirmerie de peste des Fédons à Lambesc (Bouches-du-Rhône). Bilan Archéologique du Service Régional de l'Archéologie de Provence-Alpes-Côte d'Azur, pp 857–866

Bruzek J, Schmitt A, Murail P (2005) Identification biologique individuelle en paléoanthropologie. Détermination du sexe et estimation de l'âge au décès à partir du squelette. In: Dutour O, Hublin JJ, Vandermeersch B (eds) Origine et evolution humaine. Comité des Travaux Historiques et Scientifiques, Paris, pp 217–246

Cabezuelo U, Castex D (1994) Le cimetière Saint-Pierre à Dreux (Eure-et-Loire). Dolmens, sarcophages et pierres tombales. Les pratiques funéraires en Eure-et-Loir de la préhistoire à nos jours. Maison de l'Archéologie, Chartres, pp 68–69

Castex D (1992) Étude anthropologique du cimetière paroissial Saint-Pierre, place Métezeau à Dreux (Eure-et-Loir). Rapport anthropologique, Service Régional de l'Archéologie de l'Eure-et-Loire

Castex D (1994) Mortalité, morbidité et gestion de l'espace funéraire au cours du Haut Moyen-Age. PhD Thesis, Université de Bordeaux 1, Bordeaux

Castex D (1995) Sépultures multiples: sépultures de "catastrophes"? Dossiers d'Archéologie. Les Hommes du Moyen Age 208:44–47

Castex D (1996) Etude paléodémographique du cimetière des Fédons. In: Reynaud P, Moreau N, Bouttevin C, Rigaud P, Castex D, Dutour O, Signoli M (eds) "Les Fédons" lambesc (Bouches-du-Rhône), TGV Ligne 5 – Secteur I: Avignon–Marseille. Document final de synthèse, AFAN SRA-PACA, pp 64–79

Castex D (2005) Le cimetière de peste des Fédons: des données anthropologiques aux interprétations. Des particularités démographiques propres au site des Fédons. In: Bizot B, Castex D, Reynaud P, Signoli M (eds) La saison d'une peste (Avril–Septembre 1590). Le cimetière des Fédons à Lambesc. CNRS, Paris, pp 46–53

Castex D (2007) Les anomalies démographiques: clefs d'interprétation des cimetières d'épidémies en archéologie. In: Castex D, Cartron I (eds) Epidémies et crises de mortalité du passé. Actes des séminaires (année 2005) de la Maison des Sciences de l'Homme, Ausonius Editions, Etudes 15, pp 109–138

Castex D, Cartron I (2007) Epidémies et crises de mortalité du passé, Actes des séminaires (année 2005) de la Maison des Sciences de l'Homme, Ausonius Editions, Etudes 15

Castex D, Drancourt M (2005) D'un site funéraire à la détection d'une crise épidémique. Identités biologiques et patrimoine génétique. Actes du colloque Epidémies et Sociétés dans le Monde Occidental XIe–XXIe s. Revue Sociologie Santé, Coll. "Les Etudes Hospitalières" 22:190–209

Castex D, Friess M (1998) Ein Pestsommer in der Provence. Demographische Anomalien im Kontext von Epidemien. In: Henke W, Kemkes-Grottenthaler A (eds) Epidemiologie, Frankfurt

Castex D, Réveillas H (2007) Investigation sur une crise de mortalité à Boulogne-sur-Mer (Pas-de-Calais, XVIIIe s.). Hypothèses d'interprétation. Bull Mem Soc Anthropol Paris 19:1–2

Castex D, Sellier P, Veleminsky P, Dobisikova M, Bruzek J (2003) Paleodemography and past plague epidemics: a clue for interpreting the 17th-century Svaty Benedikt cemetery in Prague. International Anthropological Congress Anthropology and Society. Memorial Congress to the 60th anniversary of death of Dr. Ales Hrdlicka, 22–24 May 2003, Praha Humpolec, Czech Republic

Castex D, Sellier P, Bruzek J, Veleminski P, Bessou M, Dobisikova M (2005) Saint-Benoît de Prague (République Tchèque, XVIIe s.): nouvelles données démographiques pour l'interprétation d'un cimetière de peste. Réunion scientifique de la Société d'Anthropologie de Paris, 1830ème Séance, 19–21 January 2005, p 227

Darmon P (1986) La longue traque de la variole, Perrin, Paris

Drancourt M, Aboudharam G, Signoli M, Dutour O, Raoult D (1998) Detection of 400-year-old *Yersinia pestis* DNA in human dental pulp: an approach to the diagnosis of ancient septicemia. Proc Natl Acad Sci USA 95:12637–12640

Drancourt M, Roux V, La-Vu D, Tran-Hung L, Castex D, Chenal-Francisque V, Ogata H, Fournier PE, Crubezy E, Raoult D (2004) Genotyping, orientalis-like *Yersinia pestis*, and plague pandemics. Emerg Infect Dis 10:1585–1592

Drancourt M, Aboudharam G, Signoli M, Raoult D, Dutour O (2005) Validation de la saisonnalité des décès et authenticité biologique de la nature de l'épidémie. L'apport de la biologie moléculaire. In: Bizot B, Castex D, Reynaud P, Signoli M (eds) La saison d'une peste (Avril–Septembre 1590). Le cimetière des Fédons à Lambesc. CNRS, Paris, pp 65–67

Duday H (2005) L'archéothanatologie. In: Dutour O, Hublin JJ, Vandermeersch B (eds) Objets et méthodes en paléoanthropologie. Comité des Travaux Historiques et Scientifiques, Paris, pp 153–214

Duday H (2006) Archaeoethnoanatology or the archaeology of death. In: Gowland R, Knüsel C (eds) Social archaeology of funerary remains. Oxbow, Oxford, pp 30–56

Faron O (1997) Le choléra à Milan en 1836, "Epidémies et Populations". In: Jacob O (ed) Annales de Démographie Historique, Paris, pp 89–114

Guignier M (1996) Les sépultures de "catastrophe" du IXe au XIe s. à Sens (Yonne). Approches historique, anthropologique et biologique. Maîtrise en Archéologie Médiévale. Université de Bourgogne, Inédit

Guignier M (1997) Les sépultures multiples du Clos des Cordeliers à Sens (Yonne). IVe-VIe s. Approches historique, anthropologique et biologique. DEA en Anthropologie, Université de Bordeaux I, Inédit

Hanakova H, Stloukal M (1988) Pohřebišttě Kolem Byvalého Kostela Svatého Benedikta V Praze. Národní Muzeum V Praze – Přirodovtědecké muzeum

Hollingsworth MF, Hollingsworth TH (1971) Plague mortality rates by age and sex in the parish of St. Botolph's without Bishopsgate, London, 1603. Pop Stud XXV:131–146

Leca AP (1982) Et le choléra s'abattit sur Paris 1832. Albin Michel, Paris

Ledermann S (1969) Nouvelles tables-types de mortalité. INED, PUF 1969, Paris (Travaux et documents, 53)

Lucenet M (1985) Les grandes pestes en France. Aubier (Floréal), Paris

Macchiarelli R, Salvadei M (1989) Early mediaval human skeletons from the thermae of Venosa, Italy. Skeletal biology and life stresses in a group presumably inhumed following an epidemic. Riv Antropol 67:105–128

Mallet E (1835) Notices sur les anciennes pestes de Genève. Ann Hyg Publique Med Leg XIV:205–221

Margerison BJ, Knüsel CJ (2002) Paleodemographic comparison of a catastrophic and an attritional death assemblage. Am J Phys Anthropol 119:134–148

Martinec V (1971) Pohřebiště při Kostele svatého Benedikta, Nálezová zpráva

Masset C (1975) La mortalité préhistorique. Cahiers du Centre de recherches préhistoriques, Université de Paris I, 4, Paris, pp 63–90

Masset C (1987) Le "recrutement" d'un ensemble funéraire. In: Duday H, Masset C (eds) Anthropologie physique et archéologie. Méthodes d'étude des sépultures. CNRS, Paris, pp 111–134

Moreau N, Duday H, Castex D, Reynaud P (2005) Le cimetière de peste des Fédons: des données anthropologiques aux interprétations. Les gestes funéraires. In: Bizot B, Castex D, Reynaud P, Signoli M (eds) La saison d'une peste (Avril–Septembre 1590). Le cimetière des Fédons à Lambesc, CNRS, Paris, pp 37–44

Poulle P (2007) Le grand Cimetière d'Issoudun (Indre) et ses sépultures multiples (fin XVIIe–début XVIIIe s.). Premiers résultats de l'étude historique et démographique. In: Castex D, Cartron I (eds) Epidémies et crises de mortalité du passé, Actes des séminaires (année 2005) de la Maison des Sciences de l'Homme. Ausonius Editions, Etudes 15, pp 77–108

Raoult D, Aboudharam G, Crubézy E, Larrouy G, Ludes B, Drancourt M (2000) Molecular identification by "suicide PCR" of *Yersinia pestis* as the agent of Medieval Black Death. Proc Natl Acad Sci USA 97:12800–12803

Razell P (1977) The conquest of smallpox: the impact of inoculation on smallpox mortality in eighteenth century Britain. Caliban Books, Firle, Sussex

Réveillas H (2005) Etude du recrutement et du fonctionnement des sépultures multiples de Boulogne-sur-Mer (XVIIIe s.), Master 2 en Anthropologie, Université de Bordeaux I, Inédit

Reynaud P, Bizot B (2005) Une épidémie de peste à Lambesc en 1590: un faisceau d'arguments archéologiques et historiques. Les faits archéologiques. In: Bizot B, Castex D, Reynaud P, Signoli M (eds) La saison d'une peste (Avril–Septembre 1590). Le cimetière des Fédons à Lambesc. CNRS, Paris, pp 11–19

Reynaud P, Moreau N, Bouttevin C, Rigaud P, Castex D, Dutour O, Signoli M (1996) "Les Fédons" lambesc (Bouches-du-Rhône), TGV Ligne 5 – Secteur I: Avignon–Marseille. Document final de synthèse, AFAN SRA-PACA

Rigaud P (2005) Une épidémie de peste à Lambesc en 1590: un faisceau d'arguments archéologiques et historiques. L'infirmerie du quartier de Farriol/les Fédons dans les sources écrites. In: Bizot B, Castex D, Reynaud P, Signoli M (eds) La saison d'une peste (Avril–Septembre 1590). Le cimetière des Fédons à Lambesc. CNRS, Paris, pp 19–35

Schmitt A (2002) Estimation de l'âge au décès des sujets adultes à partir du squelette: des raisons d'espérer, Bull Mem Soc Anthropol Paris, vol 14, no.°1–2, pp 51–73

Scott S, Duncan CJ (2001) Biology of plagues. Evidence from historical populations. Cambridge University Press

Seigne J (2007) Deux "mass burials" du VIIe siècle AP. J.-C. ou la dernière vie de l'hippodrome de Gerasa (Jerash, Jordanie). In: Castex D, Cartron I (eds) Epidémies et crises de mortalité du passé, Actes des séminaires de la Maison des Sciences de l'Homme, Ausonius Editions, Etudes 15, pp 23–38

Sellier P (1996) La mise en évidence d'anomalies démographiques et leur interprétation: population, recrutement et pratiques funéraires du tumulus de Courtesoult. In: Piningre JF (ed) Le tumulus de Courtesoult (Haute-Saône) et le Ier Age du Fer dans le bassin supérieur de la Saône. Editions de la Maison des Sciences de l'Homme (DAF), Paris, pp 188–200

Sellier P, Castex D (2001) Demographic patterns, mortality by cause and probabilities of death: is there any "anomalous demographic pattern" distinctive of plague cemeteries in archaeological samples? Peste: entre Epidémies et Sociétés, Session Poster Congrès International, 23–26 July 2001, Marseille, p 79

Signoli M (1998) Etude anthropologique de crises démographiques en contexte épidémique. aspects paléo et biodémographiques de la peste en Provence. PhD Thesis, Université de la Méditerranée

Signoli M (2005) La peste: aspects épidémiologiques actuels et passé. Comparaison avec d'autres crises démographiques et d'autres sites d'inhumation de pestiférés. In: Bizot B, Castex D, Reynaud P, Signoli M (eds) La saison d'une peste (Avril–Septembre 1590). Le cimetière des Fédons à Lambesc, CNRS, Paris, pp 72–78

Signoli M, Seguy I, Biraben JN, Dutour O (2002) Paléodémographie et démographie historique en contexte épidémique: la peste en provence au XVIIIe s. Population-F 57:821–848

Chapter 3
Archaeological Proof of an Abrupt Mortality Crisis: Simultaneous Deposit of Cadavers, Simultaneous Deaths?

Henri Duday

Abstract Several parameters have to be taken into account when considering the archaeology of death, including the number of the dead, differentiation between 'cremation' or 'incineration' and 'inhumation' and between 'primary deposits' and 'secondary deposits'. In the case of a primary deposit, the simultaneity of the deposits demonstrates *ipso facto* the simultaneity or close proximity in time of the deaths provided that there is the possibility of prolonged conservation, either by cold, desiccation, or a particular environment. In the case of secondary burials, simultaneous deposits in no way indicate simultaneous deaths. Archaeology helps demonstrate the synchronous deposition of the remains of several bodies. Dating methods are generally ineffectual in this context. In some circumstances the excavation uncovers determinative information. Biological analysis of skeletons may also provide valuable information. Finally, there remains the information from the excavation. The nature of the dead must also be taken into account. It can thus be seen that, in the absence of textual or epigraphic data, the archaeological demonstration of an abrupt mortality crisis is generally possible only when inhumations take place inside structures in which the remains of a large number of subjects are assembled within a restricted space.

3.1 Introduction

When considering how the archaeology of death, in particular funerary archaeology, can draw attention to abrupt mortality crises, the definition of terms is of primary importance (Boulestin and Duday 2005). The most obvious distinction to be made concerns the number of the dead. The death of a variable number of subjects within a relatively brief period cannot be considered in relation to an isolated individual grave (i.e. a burial place containing the remains of a single individual), but only in relation

Henri Duday
CNRS, UMR 5199 PACEA, Laboratory of Anthropology of Past Populations, Bordeaux 1 University, Avenue des Facultés, 33405 Talence cedex, France
E-mail: h.duday@anthropologie.u-bordeaux1.fr

D. Raoult and M. Drancourt (eds.), *Paleomicrobiology: Past Human Infections.*

to a complete funerary complex, which can be classified into several types. Cemeteries or necropolises assemble a number of graves (in general individual) within a defined space (sometimes delimited by a ditch, wall or fence). Each grave has its own architecture, in some cases a simple pit. Where the remains of several individuals are found within the same structure, Leclerc and Tarrête (1988) have suggested the use of the general term *sépultures plurielles* or 'plural burials'. Within this category, the term *sépultures multiples* or 'multiple burials' refers to burials where the deposition of several or many bodies is simultaneous – *sépultures doubles ou triples* or 'double or triple burials' being the *a minima* examples. On the other hand, *sépultures collectives* or 'collective burials' are the result of depositions staggered over a long period of time (often decades or, in some cases, several centuries). Obviously, this type of operation requires a system for opening and closing the funerary chamber as needed.

It is also usual to differentiate between 'cremation' or 'incineration' (treatment of the cadaver by fire) and 'inhumation' (an inappropriate term used to signify that the body has not been burned, whereas its etymology explicitly suggests the idea of placing in the ground).

Finally, as in the field of ethnology, which distinguishes simple and multiple funerals, archaeological literature contrasts 'primary deposits' (a recent cadaver, or part of a cadaver, that is still anatomically complete) with 'secondary deposits' made up of 'dry' bones no longer connected by ligaments because of decomposition or certain funerary practices (e.g. cremation).

3.2 Simultaneous Deposits, Simultaneous Deaths

In the case of a primary deposit, decomposition takes place at what becomes the final burial site of the body. As putrefaction of organic matter is a relatively rapid phenomenon, the deposition of several complete bodies implies that the subjects concerned died at or about the same time, within a period less than that necessary for the disarticulation of the first cadavers to have begun (this period is generally estimated as no more than a few weeks). It is thus considered that the simultaneity of the deposits demonstrates *ipso facto* the simultaneity or close proximity in time of the deaths.

This is a well-founded argument, provided that there has been prolonged conservation, either by cold, desiccation, a particular environment (e.g. peat bogs, anaerobic surroundings) that inhibits the action of bacteria active in the process of putrefaction, or a combination of several of these factors (for instance, the dry, very cold and well-ventilated caves found at high altitudes in the Andes). The prolongation or blocking of putrefaction may also be caused by a particular treatment of the cadaver (injections of antibacterial or fungicidal liquids, mummification). In such cases, it is of course possible to depose intact remains of subjects who died at different times in the same place, and examples of this abound (e.g. the catacombs of the Capuchin Convent in Palermo, but also morgues, or dissection rooms in medical schools). In very cold countries, when burial ditches could not be dug in the frozen ground, it was common to place the coffin containing the corpse in the snow on the roof of the house until the thaw softened the ground at the same time as the process of decomposition (until then inhibited by the low temperatures) started. If several individuals from the same community died at different times during the winter, their intact corpses, conserved by the cold, could be interred simultaneously, generally in individual graves.

In the case of secondary burials, simultaneous deposits in no way indicate simultaneous deaths. A first example concerns burial after cremation. When a cinerary urn contains the burnt bones of two individuals, their bones sometimes lie in two well-defined and superimposed layers, but nothing indicates the length of time between the two ceremonies, which may have followed closely. The bones of the two individuals can, on the other hand, be completely mixed. In this case there is no proof that they were burned together, or, of course, that they died at the same time; it is perfectly feasible to burn a corpse, to keep the remains in a temporary receptacle until the death of the individual with whom they are to be associated, to burn this individual and, finally, to mix the bones of the two individuals in the same cinerary urn. Another example is that of Neolithic collective burials, where it is common to find the cranio-facial blocks and the long bones of the arms and legs arranged along the walls of the funerary chamber (dolmen, sepulchre or hypogeum). These practices, which take place after decomposition of the bodies, often concern the detached bones of a number of individuals; nothing leads us to suppose that they were put in place at the same time, nor *a fortiori* that the subjects died at the same time.

3.3 Demonstration of the Simultaneity of Deposits

If a relationship between simultaneous deposits and simultaneous deaths is envisaged, it remains to be seen how archaeology can demonstrate the synchronous deposition of the remains of several bodies. Conditions vary according to funerary practices and treatments.

Dating methods are generally ineffectual in this context, and do not allow precision on the order of days or weeks, either in the case of absolute dates (methods using physics or chemistry), or in that of relative dates (chronology of the different elements of associated accessories or equipment). Dendrochronology allows dating to within a year, or even a season, of course, but the chronological link between the felling of a tree and its use in a funerary context (built elements, coffin) must also be established.

In some, quite exceptional, circumstances the stratigraphy can be a determining contribution: this is the case when bodies have been suddenly buried by a mudslide, landslip, collapse of a wall or building (earthquake), or by volcanic ash (Pompeii or Herculanum). However, these are natural catastrophes outside the funerary context.

In some circumstances, the excavation uncovers determinative information, for example, epitaphs where the date of death is explicitly indicated or commemorative inscriptions that relate a particular event (as is sometimes the case on battlegrounds). In historical periods, records may relate an abrupt mortality crisis and indicate the funerary site (e.g. Les Fédons, Lambesc) (Bizot et al. 2005; Moreau et al. 2005).

Biological analysis of the skeletons may also provide valuable information by showing that all the deaths at a given site have the same cause. This approach has been restricted for a long time to warlike activity (fatal injuries from weapons), but it is now possible in the study of epidemics thanks to the progress of molecular palaeobiochemistry (identification of the DNA of infectious agents, in particular *Yersinia pestis*). These methods are, of course, very costly and cannot be employed indiscriminately. It is necessary to restrict their use to funerary sites at which there are serious indications of a mortality crisis unconnected to a massacre or act of war, either by the very character of the deposit (cf. *infra*) or by peculiarities of the mortality curve [in such cases, the anomalies detected in the distribution by age group should reveal an 'unnatural' mortality (compared to a mortality outside a period of crisis) and not a selection biased for cultural reasons].

Finally, there remains the information from the excavation. Archaeothanatology is totally ineffectual when the skeletons are not in direct contact with each other, for example, in cemeteries and necropolises, or in plural burials where the number of bodies is very low with regard to the available area (see Chambon 2003, about the tumulus 'La Hoguette' at Fontenay-le-Marmion). On the contrary, when several bodies are found in a restricted space the relative chronology of the articular dislocations may be used. If the deposits are staggered in time, the laying down of a new subject will inevitably perturb the arrangement of the skeletons already present; secondary gestures of 'reduction' (which do not correspond to true secondary burials) are frequently observed. However, if the deposits are simultaneous, the (articular) connections should be more strictly respected, because all the bodies will decompose at the same time; the displacements observed result from the action of gravity [with the exception of possible ulterior rearrangements (anthropic intervention, burrowing animals, water drip, collapse of structures…)], bones liberated by decomposition slip into the spaces freed by the disappearance of the soft tissue of the subjects underneath – these are principally vertical displacements.

This method is obviously much more effective than the usual dating methods employed in archaeology, the limits of discrimination being fixed to the time necessary for the destruction of the most labile articular connections (those loosened most rapidly during decomposition) (Duday 2005a, 2005b, 2006). This period is, however, of the order of a few weeks (it varies considerably according to climatic conditions and, naturally, funerary treatments) and thus it is not possible to differentiate between truly simultaneous deposits and those separated by a few days or weeks. Under some circumstances, this is not important as an abrupt mortality crisis is defined precisely as the death of a relatively large number of subjects within a relatively short period of time.

The nature of the dead must also be taken into account. The simultaneous death of several members of the same family, either in a traffic accident, poisoned by a dish of *amanita phalloides* (Death Caps), or intoxicated by carbon monoxide from a faulty boiler, while certainly a dramatic event in a household, represents no more than a news item in the town in which they live. In such cases, nobody would suggest an abrupt mortality crisis, and accidents of this type must be considered when a double or triple burial is uncovered during an excavation, as well as the reasons for an inhumation at the same time within the same structure (this question can be profitably referred to in the remarkable work of A. Testart, dealing with 'associated deaths') (Testart 2004).

It can thus be seen that, in the absence of textual or epigraphic data, the archaeological demonstration of an abrupt mortality crisis is generally possible only when inhumations take place inside structures in which the remains of a large number of subjects are assembled within a restricted space. Although we can sometimes (more and more often) specify the relative chronology of the deposits in secondary burials, whether cremations or inhumations, we can in no way indicate the moment of death. As far as cemeteries and necropolises are concerned, it is very difficult to place the tombs on a timescale if they are dissociated, or even if they are adjacent or aligned; the information given by the possible intersection of pits and associated accessories or equipment is too imprecise to guarantee the necessary discrimination (of the order of a few days to a few weeks). It is therefore evident that, although the archaeology of death has made enormous progress in the study of these very particular funerary assemblages, methods for the recognition of all deposits that may be due to abrupt mortality crises are inadequate as yet.

It would seem that the use of large burial pits represents only one of the modalities – the most spectacular, but certainly not the only form – of management of cadavers in such contexts (the cemetery at Les Fédons is an excellent example). The archives indicate a connection with an epidemic of plague, and molecular palaeomicrobiochemistry has revealed the DNA of *Yersinia pestis*, but nevertheless most of the dead were deposited in individual pits. This may have resulted from urgent inhumation, perhaps when the rate of death was too rapid for gravediggers to bury each cadaver individually. They represent, nonetheless, a veritable funerary treatment [Thus military archives indicate that the pit at Saint-Remi-la-Calonne, where the German army inhumated the bodies of 21 French officers and soldiers (including that of the author Alain Fournier) killed on 22nd September 1914 at the front, represents without any doubt a true grave] (Adam et al. 1993) and in this sense provide fundamental cultural information. It is for this reason that comparison with the management of the cadavers at 'normal' times (outwith periods of crisis) is seen as a priority in archaeological funerary research.

References

Adam F, Boura F, Duday H (1993) La fouille de Saint-Rémi-la-Calonne: une opération d'archéologie funéraire expérimentale, ou l'anthropologie de terrain en quête de ses références. Les Nouvelles de l'Archéologie, 48/49, pp 59–61

Bizot B, Castex D, Reynaud P, Signoli M (eds) (2005) La saison d'une peste (Avril–Septembre 1590). Le cimetière des Fédons à Lambesc. CNRS, Paris

Boulestin B, Duday H (2005) Ethnologie et archéologie de la mort: de l'illusion des references à l'emploi d'un vocabulaire. In: Mordant C, Depierre G (eds) Les pratiques funéraires à l'âge de Bronze en France. Actes de la table ronde de Sens-en-Bougogne, Yonne, pp 17–35

Chambon P (2003) Les morts dans les sépultures collectives néolithiques en France. Du cadavre aux restes ultimes. XXXV Supplément à Gallia Préhistoire, CNRS, Paris

Duday H (2005a) Lezioni di Archeotanatologia. Archeologia funeraria e anthropologica di campo. Soprintendenza archeologica di Roma et Ecole Française de Rome, Rome

Duday H (2005b) L'archéothanatologie ou l'archéologie de la Mort. In: Dutour O, Hublin J-J, Vandermeersch B (eds) Objets et méthodes en Paléoanthropologie. Comité des Travaux Historiques et Scientifiques, Paris, pp 153–215

Duday H (2006) Archeoethnoanatology or the Archeology of Death. In: Gowland R, Knüsel C (eds) Social archeology of funerary remains. Oxbow, Oxford, pp 30–56

Leclerc J, Tarrête J (1988) Sépulture. In: Leroi-Gourhan A (ed) Dictionnaire de la Préhistoire. Presses Universitaires de France, Paris, pp 963–964

Moreau N, Duday H, Castex D, Reynaud P (2005) Le cimetière de peste des Fédons: des données anthropologiques aux interprétations. Les gestes funéraires. Description de sépultures représentatives. In: Bizot B, Castex D, Reynaud P, Signoli M (eds) La saison d'une peste (Avril–Septembre 1590). Le cimetière des Fédons à Lambesc. CNRS, Paris, pp 37–44

Testart A (2004) Les morts d'accompagnement. La servitude volontaire, I. Editions Errance, Paris

Chapter 4
Molecular Detection of Past Pathogens

Michel Drancourt(✉) and Didier Raoult

Abstract Detection and characterisation of DNA is the most widely used approach for the study of past pathogens. This approach can be applied to various specimens, including environmental, vector and animal reservoir specimens as well as human corpses. Experimental data indicated that host-associated microbial DNA can survive for 20,000 years, and bacterial DNA preserved in permafrost specimens has been dated up to 1 million years. Current protocols targeted one pathogen at a time and universal 16S rDNA-based detection of bacteria have yielded ambiguous results. There is no universal detection of ancient virus so far. Major human pathogens, e.g. *Mycobacterium tuberculosis*, *Mycobacterium leprae*, *Yersinia pestis*, *Rickettsia prowazekii*, *Bartonella* spp. and Spanish influenza virus have been detected in suitable human specimens. Ancient *M. tuberculosis* and *Y. pestis* organisms have been genotyped, whereas the entire RNA genome of Spanish influenza virus was reconstituted for extensive studies. Metagenomic approaches based on high throughput pyrosequencing may help further resolve forthcoming issues. Interpretation of experimental data has to be based upon strict rules due to potential contamination of specimens.

4.1 Introduction

As a discipline, palaeomicrobiology (Zink et al. 2002) began in 1993 with the molecular detection of *Mycobacterium tuberculosis* DNA in an ancient human skeleton (Spigelman and Lemma 1993). This finding served to illustrate the importance of molecular biology techniques in the quest for pathogens, and microbes at large, in ancient specimens recovered from various human tissues, as well as from environmental samples of potential vectors and reservoirs of past pathogens

Michel Drancourt
Unité des Rickettsies, CNRS UMR 6020, Faculté de Médecine, Université de la Méditerranée, 27 Boulevard Jean Moulin, 13385 Marseille Cedex 5, France
E-mail: Michel.Drancourtt@medecine.univ-mrs.fr

D. Raoult and M. Drancourt (eds.), *Paleomicrobiology: Past Human Infections.*

(Drancourt et al. 2005). Indeed, with the exception of some enteric parasites (Bouchet et al. 2003) and rare human viruses, all past pathogens and microbes have been detected and studied thanks to the detection and characterisation of nucleic acids. Experimental data have now demonstrated that bacterial DNA can be detected in 20,000-year-old host specimens, and in up to several thousand-year-old environmental specimens preserved in permafrost (Willerslev et al. 2005). Likewise, Spanish influenza virus RNA has been extensively studied after its recovery from both formalin-preserved human lung tissue (Reid et al. 2000; Taubenberger et al. 1997) and permafrost-preserved human tissues (Reid et al. 2000).

The objectives of molecular detection of past pathogens include the diagnosis of past infectious diseases through the identification of specific molecular sequences in ancient remains; the elucidation of the epidemiology of past infectious diseases by reconstituting the temporal and geographical distribution of infected individuals, reservoirs and vectors; and the tracing of the genetic evolution of the microorganisms themselves through genotyping (Drancourt and Raoult 2005). Data from such studies benefit modern microbiology and studies of host–pathogen relationships. Refinements in molecular typing now allow researchers to study the genetic evolution of microorganisms and the timing of their introduction into human populations. Initial palaeomicrobiological studies used bone tissue, whereas later studies have progressed to using mummified tissues and dental pulp for analysis (Salo et al. 1994; Drancourt et al. 1998; Raoult et al. 2000) (Fig. 4.1). As for bone tissues, it was shown that both the gross and histological preservation were correlated with DNA survival (Haynes et al. 1970). Concomitantly, experimental standards for palaeomicrobiology have emerged to deal with the problems of contamination and the authenticity of data.

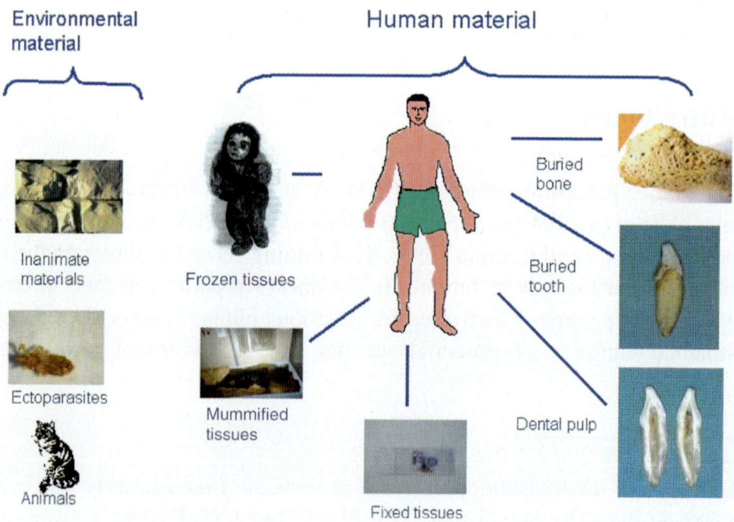

Fig. 4.1 Suitable source materials for amplification and sequencing of ancient microbe DNA

4.2 Protocols for the Molecular Detection of Past Pathogens

Detection and identification of pathogens in ancient human and environmental specimens relies mostly upon the molecular detection of specific nucleic acid sequences. A few studies have focussed on viral RNA for the detection of Spanish influenza virus (Reid et al. 2000; Taubenberger et al. 1997; Reid et al. 1999) but the vast majority of studies have targeted ancient bacterial and parasite DNA. While experimental protocols for DNA extraction and its amplification by polymerase chain reaction (PCR) have been empirical, a few systematic studies of experimental parameters now provide clear experimental guidelines for optimal DNA extraction and amplification from bone tissues (Rohland and Hofreiter 2007).

4.2.1 Ancient DNA Characteristics

Empirical observations made over the last 20 years indicated that ancient DNA has adverse characteristics when compared to modern DNA. The amino-acid racemisation ratio was shown to predict the preservation of ancient DNA (Poinar et al. 1996). Ancient DNA is broken into pieces of <500 bp (Lindahl 1993); consequently, PCR cannot be used to amplify large fragments in ancient specimens. In the case of ancient mammal DNA, this limitation has been circumvented by pre-treatment of the ancient DNA with reconstructive polymerisation (Golenberg et al. 1996) or enzymatic repair by the combined activities of DNA polymerase I and T4 DNA ligase (Pusch et al. 1998; Di et al. 2002). However, nothing has been published regarding the repair of ancient microbial DNA.

A second feature of ancient DNA is chemical modification, comprising both oxidisation and hydrolysis resulting in deamination of nucleotides (Hoss et al. 1996; Hofreiter et al. 2001). Such modifications have been implicated in cases of poor yields from PCR. It has been recently demonstrated that not all DNA polymerases amplify ancient DNA extracted from cave bear bone with the same efficiency (Rohland and Hofreiter 2007).

Third, numerous studies have demonstrated the presence of poorly characterised PCR inhibitors in ancient specimen extracts (Hoss et al. 1996; Hanni et al. 1995). The precise nature of these inhibitors, once correlated to the presence of a brown coloration of extracts (Hanni et al. 1995), is not known. Two strategies have been proposed to circumvent the presence of inhibitors: dilution of extracted specimens and the addition of bovine serum albumin (BSA). The effectiveness of both solutions has recently been demonstrated (Rohland and Hofreiter 2007).

4.2.2 Nucleic Acid Extraction

Since the initial demonstration that DNA can survive in mummified human tissues (Pääbo 1985), nucleic acid extraction from various types of specimens has been

Table 4.1 Adverse characteristics of ancient microbial DNA limiting PCR-based detection of past pathogens and proposed solutions. *PCR* Polymerase chain reaction, *BSA* bovine serum albumin

Characteristic	Consequence for PCR-based detection	Proposed solutions
Fragmentation <500 bp	Amplification of small fragments only	Select PCR primers in order to amplify a fragment ≤300 bp DNA enzymatic repair using DNA polymerase I/T4 DNA ligase[a]
Chemical alterations	Poor PCR yield	Select appropriate *Taq* DNA polymerase
PCR inhibitors	Lack of PCR amplification	Run dilutions of extracted DNA Add BSA to PCR mix

[a]This technique has been published only for ancient eukaryotic DNA

achieved. Extraction can be achieved from conjunctive tissues that have been either frozen (Reid et al. 1999; Cano et al. 2000; Rhodes et al. 1998), mummified (Salo et al. 1994; Fornaciari et al. 2003) buried (Reid et al. 1999) (Table 4.1) or fixed (Taubenberger et al. 2005) . Extraction from bone tissues requires extensive decalcification using EDTA and mechanical grinding prior to DNA extraction. The same holds true for entire teeth. We proposed the use of dental pulp as a suitable specimen for the molecular detection of blood-borne organisms (Drancourt et al. 1998). Several protocols for the extraction of DNA from ancient tissues have been proposed, but the comparative performance of these various protocols has been evaluated only recently (Rohland and Hofreiter 2007).

4.2.3 Amplification, Cloning and Sequencing

All studies dealing with ancient microbial DNA use a PCR amplification step before nucleotide sequencing. Various PCR protocols have been developed, including one-step conventional PCR in most studies, nested and hemi-nested PCR and, rarely, real-time PCR. The addition of either BSA or a related protein in the PCR cocktail had been advocated in order to prevent PCR inhibition (Rohland and Hofreiter 2007). This empirical observation has recently been verified (Rohland and Hofreiter 2007). The exact nature of the PCR inhibitors in ancient specimens has not been elucidated, and the proposed correlation of the brown colour of the extraction product with PCR inhibition (Hanni et al. 1995) has not been confirmed (Drancourt et al. 1998). In most studies, PCR-amplified fragments are cloned before being sequenced. So far, the conventional Sanger sequencing method has been applied using capillary automatic sequencers.

4.3 Contamination of Ancient Specimens

Micro-organisms from the burial site can contaminate specimens before laboratory analyses, whereas laboratory micro-organisms and their DNA can contaminate specimens during laboratory analyses. Some PCR mix reagents, including PCR primers, polymerases and water used to complement reaction volumes, have been shown to be contaminated by bacterial DNA. In the detection of past bacteria, the contamination threat is particularly great when using a universal approach such as 16S rDNA-based PCR (Gilbert et al. 2004; Zink et al. 2000; Cano et al. 2000). Specific molecular targets carry a smaller risk. The specificity of detection has been shown by analysis of environmental samples in parallel with buried specimens (Papagrigorakis et al. 2006). The use of naturally protected specimens, such as dental pulp, might also limit the risk of external contamination (Drancourt et al. 1998).

4.4 Strategies to Obtain Reliable Data

Several protocols can be used to limit the risk of contamination in the laboratory (Fig. 4.2, Table 4.2). The external cleansing of bone using filtered compressed air and sterile distilled water, scraping the external surface, and irradiation with 254-nm ultraviolet (UV) light have all been advocated (Ou et al. 1991). For the manipulation of ancient teeth, encasing the specimen in sterile resin has been proposed (Gilbert et al. 2003). All PCR-based experiments should be carried out in designated one-way PCR suites with appropriate ventilation. Primer optimisation for PCR should be carried out in a separate building from the one in which the ancient material is handled, and

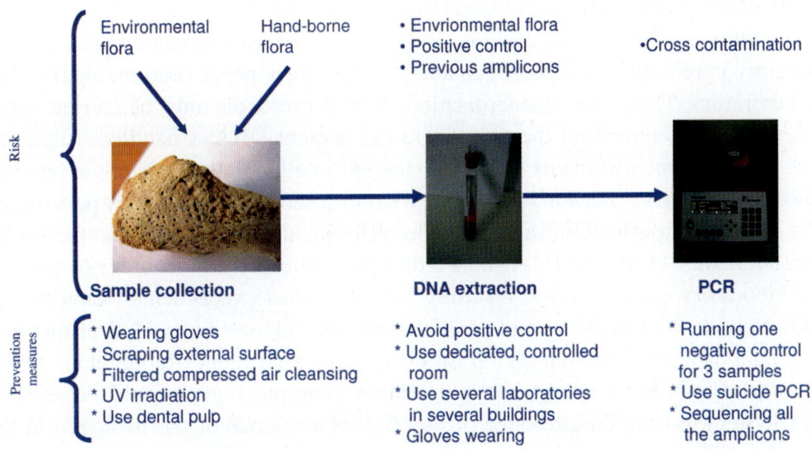

Fig. 4.2 Prevention of bacterial and molecular contamination in palaeomicrobiology

Table 4.2 Prevention of specimen contamination in ancient microbial DNA studies

Source of contamination	Proposed solutions
Burial site:	
Environmental flora	External surface scraping, sterile water and filtered compressed air cleansing
	UV irradiation
	Using dental pulp
Hand-borne flora	Wearing gloves for specimen manipulation
Laboratory:	
Environmental flora	Wearing gloves for specimen manipulation
Hand-borne flora	Respect of strict protocols
PCR reagents	Use of dedicated, controlled rooms
Previous experiments	Suicide PCR
Cross-contamination	No positive control
	Negative controls
	Amplicon sequencing

PCR and post-PCR experiments should be performed in a separate room using disposable equipment and freshly prepared reagents that have been irradiated with UV light. It has been also advocated that ancient DNA experiments be performed without using a positive control. Alternatively, mock positive controls and DNA from different, related species can be used. Furthermore, we developed "suicide PCR" reactions, which target a new genomic region by using a new PCR primer pair in every new experiment, to prevent vertical contamination from previous amplifications (Raoult et al. 2000). The introduction of numerous negative controls helps monitor any carry-over source of contamination. Material collected from unaffected individuals are also of value; for example, lesion-free bones collected from fossilised *Canis* and *Equus* species have been used as controls for the molecular detection of *M. tuberculosis* DNA in extinct bison (Rothschild et al. 2001).

As pathogens are not ubiquitous organisms, the first sequence achieved in a laboratory is reliable if the pathogen and its DNA have never been manipulated in that laboratory. Therefore, standardisation of PCR protocols must be carried out in a laboratory different from the one where the ancient DNA is handled. Likewise, DNA from ancient specimens must be extracted in a laboratory where the targeted pathogens have never been manipulated. We optimised this approach by performing these different experimental steps in laboratories located in different campus buildings (Drancourt et al. 2004). Also, we designed suicide PCR in order to prevent intra-laboratory contamination resulting from previous experiments (Raoult et al. 2000). Suicide PCR avoids use of positive controls and uses a new PCR primer pair targeting a different genomic region for every new experiment (Raoult et al. 2000). Alternatively, PCR targetting a hypervariable genomic region could be used in order to demonstrate the presence of an original sequence of the pathogen in the ancient specimen.

4.5 Interpretation of Data

Strict adherence to the rules for the prevention of contamination is a first step towards ensuring the authenticity of ancient microbial isolates. Absence of any amplicon in negative controls is strictly required. The recovery of an original sequence indicates that laboratory contamination has not occurred and is good evidence for authenticity. The original sequence must be shown in several clones. Chemical modifications of ancient DNA can result in "jumping PCR" – template switching during PCR and C→T and G→A substitutions. The sequencing of multiple clones derived from more than one independent amplification has been advocated to reduce the risk of obtaining incorrect DNA sequences (Hoss et al. 1996; Spencer and Howe 2004). However, there is no evidence of "spontaneous" mutation in ancient DNA (Serre et al. 2004).

Phylogenetic analyses of the gene sequence from the ancient microorganism can confirm its antiquity; for example, phylogenetic analyses of a *Bacillus* sp. that was once claimed to be 250 million years old showed that it was in fact a modern contaminant (Vreeland et al. 2000; Nickle et al. 2002). The reproducibility of results using different specimens collected from the same individual is another a criterion. Also, the demonstration of two unrelated sequences that identify the same pathogen in the same specimen further increases the specificity of the identification.

4.6 Molecular Detection of Past Pathogens: Current Data

Most published data deal with the detection and molecular characterisation of ancient bacteria, while fewer studies have examined past viruses and parasites. The most significant data are presented in Table 4.3. To complement the molecular detection and identification of past pathogens, some ancient bacteria have been genotyped. In the case of the *M. tuberculosis* complex, ancient mycobacteria were genotyped by sequencing the phospholipase-C *mtp*40 gene, a *Mycobacterium tuberculosis*-specific region, another *Mycobacterium bovis*-specific fragment and the *oxy*R pseudogene (Pääbo 1985). This work demonstrated that medieval mycobacteria were more closely related to modern *Mycobacterium tuberculosis* than to *Mycobacterium bovis*. Similar conclusions were obtained from a spoligotyping analysis of 12 *Mycobacterium tuberculosis* strains that were characterised among Egyptian mummies dating from 2050 to 500 B.C. (Zink et al. 2003). Spoligotypes obtained from mycobacterial DNA from an extinct bison demonstrated that it was more closely related to the *Mycobacterium tuberculosis* / *Mycobacterium africanum* group than it was to *M. bovis* (Rothschild et al. 2001). These data indicated that the theory that *Mycobacterium tuberculosis* had evolved from *M. bovis* by specific adaptation to the human host was not in fact the case (Stead et al. 1995).

In our laboratory, using multispacer sequence typing (MST), we have successfully genotyped *Yersinia pestis* in individuals suspected to have died from the Justinian

Table 4.3 Current data in palaeomicrobiology. *BP* before present (years)

Bacteria	Source	Specimen, body site	Conservation	Date	Reference
Mycobacterium tuberculosis	Bison	Metacarpal	Buried	17,000 BP	Rothschild et al. 2001
	Human	Lung, lymph node	Mummified	1,000 BP	Salo et al. 1994
	Human	Bone	Mummified	5,400 BP	Crubezy et al. 1998
	Human	Metacarpal, lumbar vertebrae	Buried	Medieval	Taylor et al. 1999
	Human	Rib	Buried	Medieval	Mays et al. 2002
	Human	Vertebrae	Buried	1,000 BP	Arriaza et al. 1995
	Human	Mandible	Buried	1400–1800 A.D.	Hass et al. 2000a
	Human	Vertebrae, femur, ankle, rib, pleura	Buried	Seventh–eighth centuries; seventeenth century	Hass et al. 2000a
	Human	Lung pleura	Buried	600 A.D.	Donoghue et al. 1998
	Human	Bone	Buried	1,000 BP	Gernaey et al. 2001
	Human	Vertebrae, rib	Buried	400–230 B.C.	Mays and Taylor 2003
	Human	Bone, soft tissues	Mummified	2050–500 B.C.	Hanni et al. 1995
	Human	Bone	Buried		Spigelman and Lemma 1993
	Human	Lungs, pleura, abdomen, ribs, hair, teeth	Mummified	Eighteenth–nineteenth centuries	Fletcher et al. 2003
	Human	Wrist, lumbar vertebrae	Buried	Fourteenth–sixteenth centuries	Taylor et al. 1996
Mycobacterium leprae	Human	Foot bones	Buried	Twelfth century	Montiel et al. 2003
	Human	Metacarpals	Buried	300–600 A.D.	Spigelman and Donoghue 2001
	Human	Skulls	Buried	1400–1800 A.D.	Donoghue et al. 2001
	Human	Hard palate, skull	Buried	1400–1800 A.D.; tenth century	Hass et al. 2000a
Enteric bacteria	Mastodon	Bowel	Frozen	12,000 BP	Rhodes et al. 1998
	Human	Metatarse	Mummified	1400 B.C.	Zink et al. 2000

	Human	Upper gut content	Preserved in bog	300 B.C.	Fricker et al. 1997
Treponema pallidum	Human	Bone	Buried	240 BP	Kolman et al. 1999
Borrelia burgdorferi	Ticks		Dry	1884	Matuschka et al. 1996
	Rodents		Dry	Nineteenth century	Marshall et al. 1994
Spirochetes	Termite	Instestinal tissue	Amber	Miocene	Wier et al. 2002
Bartonella quintana	Human	Dental pulp	Buried	4,000 BP	Drancourt et al. 2005
Bartonella henselae	Cat	Dental pulp	Buried	Thirteenth–eighteenth centuries	La et al. 2004
	Human	Dental pulp	Buried	Fifth–fourteenth centuries	Drancourt et al. 2004
	Human	Dental pulp	Buried	1590–1722	Drancourt et al. 1998
	Human	Dental pulp	Buried	1348	Raoult et al. 2000
Rickettsia prowazekii	Human	Dental pulp	Buried	1812	Raoult et al. 2006
Mixed flora	Human	Skin/muscle	Frozen	Neolithic	Rollo et al. 2000
Mixed flora	Human	Colon	Frozen	Neolithic	Cano et al. 2000
Parasites					
Ascaris lumbricoïdes	Human	Coprolites	Buried	Middle-Ages	Loreille et al. 2001
Plasmodium falciparum	Human	Bone	Burried	1,500 BP	Taylor et al. 1997
Trypanosoma cruzi	Human	Human, visceral tissue	Mummified	4,000 BP	Guhl et al. 1997
	Human	Heart, lung, liver, kidney, ileum; colon, muscle, brain	Mummified	9,000 BP	Aufderheide et al. 2004
	Human	Bone	Mummified	4,000 BP	Zink et al. 2006
Enterobius vermicularis	Human	Corpolites	Buried		Loreille et al. 2001
Viruses					
HTLV-I	Human	Bone		1,500 BP	Li et al. 1999
HPV	Human	Skin	Mummified	Sixteenth century	Fornaciari et al. 2003
Influenza virus	Human	Lung	Fixed	1918	Taubenberger et al. 1997
			Frozen	1918	Reid et al. 2000

plague (Drancourt et al. 2004). After comparison of the two *Y. pestis* genome sequences available in GenBank, we found that some intergenic spacer sequences were highly variable, and we amplified six of these sequences from the ancient specimens. Sequence analyses showed that the sequences obtained were original sequences owing to the presence of point mutations. These mutations were consistently found in several clones, therefore confirming that they were not merely caused by misincorporation of nucleotides by *Taq* polymerase. *Y. pestis* has been subdivided into three biovars on the basis of their ability to ferment glycerol and to reduce nitrate. On the basis of their current geographical niche, and on historical records that indicated the geographical origin of the pandemics, it was speculated that each biovar caused a specific pandemic (Devignat 1954). MST data demonstrated that the genotype involved in all three pandemics was associated with the Orientalis biovar, a result recently confirmed by demonstration of a specific deletion in the *glp*D gene (Drancourt et al. 2007).

4.7 Future Research

The detection of pathogens in their ancient reservoirs, and of vectors, will be a key factor in achieving the goal of a global epidemiology scheme for every transmissible infectious disease. Such detection will benefit from improved collaboration between palaeozoologists, specialists in ancient ectoparasites and palaeomicrobiologists. Specific issues include the correct collection and identification of buried animals and ectoparasites. With regards to human remains and the remains of other mammals, in our opinion, the broad use of dental pulp could help resolve the aetiology of ancient bloodborne infections, although universal protocols are still required.

The application of the universal 16S rDNA-based detection and identification of bacteria to palaeomicrobiology has been limited by contamination of the ancient material. However, this powerful molecular tool will be invaluable in the study of the nature and epidemiology of unpredicted pathogens. The aetiology of numerous past epidemics remains unknown, despite testing for the presence of one or more bacterial pathogens. Tracing any bacterial pathogen within the remains of this past population could help resolve the question of the aetiology of some mysterious epidemics. Given the small amount of material available in the majority of these cases, testing for all bacterial pathogens simultaneously would be helpful. Studies must be performed to develop a protocol of universal amplification and sequencing that is adapted to ancient bacterial DNA.

Metagenomic analysis of total DNA extracted from ancient specimens is a promising field of research. It relies on the high throughput sequencing made possible by the new generation of pyrosequencers. This new approach has been successfully applied to the study of complex modern flora, and to that of ancient mammoth tissue (Poinar et al. 2006). It may resolve the quest for universal detection, not only of bacteria but also of viruses, in ancient specimens.

Genotyping will create the necessary bridge between the detection of microbial DNA in ancient environmental and human specimens and modern microbiology. The availability of a large database of complete microbial genome sequences has already prompted the establishment of suicide PCR and new genotyping methods for past microorganisms, including spoligotyping of *M. tuberculosis* (Zink et al. 2003) and MST of *Y. pestis* (Drancourt et al. 2004). Such efforts should be continued.

References

Arriaza BT, Salo W, Aufderheide AC, Holcomb TA (1995) Pre-Columbian tuberculosis in northern Chile: molecular and skeletal evidence. Am J Phys Anthropol 98:37–45

Aufderheide AC, Salo W, Madden M, Streitz J, Buikstra J, Guhl F, Arriaza B, Renier C, Wittmers LE Jr, Fornaciari G, Allison M (2004) A 9,000-year record of Chagas' disease. Proc Natl Acad Sci USA 101:2034–2039

Bouchet F, Harter S, Le BM (2003) The state of the art of paleoparasitological research in the Old World. Mem Inst Oswaldo Cruz 98[Suppl 1]:95–101

Cano RJ, Tiefenbrunner F, Ubaldi M, Del CC, Luciani S, Cox T, Orkand P, Künzel KH, Rollo F (2000) Sequence analysis of bacterial DNA in the colon and stomach of the Tyrolean Iceman. Am J Phys Anthropol 112:297–309

Crubezy E, Ludes B, Poveda JD, Clayton J, Crouau-Roy B, Montagnon D (1998) Identification of Mycobacterium DNA in an Egyptian Pott's disease of 5,400 years old. C R Acad Sci III 321:941–951

Devignat R (1954) Biological and biochemical behavior of *Pasteurella pestis* and *Pasteurella pseudotuberculosis*. Bull World Health Org 10:463–494

Di BG, Del GS, Cammarota M, Galderisi U, Cascino A, Cipollaro M (2002) Enzymatic repair of selected cross-linked homoduplex molecules enhances nuclear gene rescue from Pompeii and Herculaneum remains. Nucleic Acids Res 30:e16

Donoghue HD, Spigelman M, Zias J, Gernaey-Child AM, Minnikin DE (1998) Mycobacterium tuberculosis complex DNA in calcified pleura from remains 1400 years old. Lett Appl Microbiol 27:265–269

Donoghue HD, Holton J, Spigelman M (2001) PCR primers that can detect low levels of *Mycobacterium leprae* DNA. J Med Microbiol 50:177–182

Drancourt M, Raoult D (2005) Palaeomicrobiology: current issues and perspectives. Nat Rev Microbiol 3:23–35

Drancourt M, Aboudharam G, Signoli M, Dutour O, Raoult D (1998) Detection of 400-year-old *Yersinia pestis* DNA in human dental pulp: an approach to the diagnosis of ancient septicemia. Proc Natl Acad Sci USA 95:12637–12640

Drancourt M, Roux V, Dang LV, Tran-Hung L, Castex D, Chenal-Francisque V, Ogata H, Fournier P-E, Crubezy E, Raoult D (2004) Genotyping, Orientalis-like *Yersinia pestis*, and plague pandemics. Emerg Infect Dis 10:1585–1592

Drancourt M, Tran-Hung L, Courtin J, Lumley H, Raoult D (2005) *Bartonella quintana* in a 4000-year-old human tooth. J Infect Dis 191:607–611

Drancourt M, Signoli M, Dang LV, Bizot B, Roux V, Tzortzis S, Raoult D (2007) *Yersinia pestis orientalis* in remains of ancient plague patients. Emerg Infect Dis 13:332–333

Fletcher HA, Donoghue HD, Holton J, Pap I, Spigelman M (2003) Widespread occurrence of Mycobacterium tuberculosis DNA from 18th–19th century Hungarians. Am J Phys Anthropol 120:144–152

Fornaciari G, Zavaglia K, Giusti L, Vultaggio C, Ciranni R (2003) Human papillomavirus in a 16th century mummy. Lancet 362:1160

Fricker EJ, Spigelman M, Fricker CR (1997) The detection of *Escherichia coli* DNA in the ancient remains of Lindow Man using the polymerase chain reaction. Lett Appl Microbiol 24:351–354

Gernaey AM, Minnikin DE, Copley MS, Dixon RA, Middleton JC, Roberts CA (2001) Mycolic acids and ancient DNA confirm an osteological diagnosis of tuberculosis. Tuberculosis 81:259–265

Gilbert MT, Willerslev E, Hansen AJ, Barnes I, Rudbeck L, Lynnerup N, Cooper A (2003) Distribution patterns of postmortem damage in human mitochondrial DNA. Am J Hum Genet 72:32–47

Gilbert MT, Cuccui J, White W, Lynnerup N, Titball RW, Cooper A, Prentice MB (2004) Absence of *Yersinia pestis*-specific DNA in human teeth from five European excavations of putative plague victims. Microbiology 150:341–354

Golenberg EM, Bickel A, Weihs P (1996) Effect of highly fragmented DNA on PCR. Nucleic Acids Res 24:5026–5033

Guhl F, Jaramillo C, Yockteng R, Vallejo GA, Cardenas-Arroyo F (1997) *Trypanosoma cruzi* DNA in human mummies. Lancet 349:1370

Hanni C, Brousseau T, Laudet V, Stehelin D (1995) Isopropanol precipitation removes PCR inhibitors from ancient bone extracts. Nucleic Acids Res 23:881–882

Hass CJ, Zink A, Molar E, Szeimies U, Reischl U, Marcsik A, Ardagna Y, Dutour O, Pálfi G, Nerlich AG (2000a) Molecular evidence for different stages of tuberculosis in ancient bone samples from Hungary. Am J Phys Anthropol 2000:293–304

Hass CJ, Zink A, Palfi G, Szeimies U, Nerlich AG (2000b) Detection of leprosy in ancient human skeletal remains by molecular identification of *Mycobacterium leprae*. Am J Clin Pathol 2000:428–436

Haynes RE, Sanders DY, Cramblett HG (1970) Rocky Mountain spotted fever in children. J Pediatr 76:685–693

Hofreiter M, Jaenicke V, Serre D, Haeseler AA, Pääbo S (2001) DNA sequences from multiple amplifications reveal artifacts induced by cytosine deamination in ancient DNA. Nucleic Acids Res 29:4793–4799

Hoss M, Jaruga P, Zastawny TH, Dizdaroglu M, Pääbo S (1996) DNA damage and DNA sequence retrieval from ancient tissues. Nucleic Acids Res 24:1304–1307

Kolman CJ, Centurion-Lara A, Lukehart SA, Owsley DW, Tuross N (1999) Identification of *Treponema pallidum* subspecies *pallidum* in a 200-year-old skeletal specimen. J Infect Dis 180:2060–2063

La VD, Clavel B, Lepetz S, Aboudharam G, Raoult D, Drancourt M (2004) Molecular detection of *Bartonella henselae* DNA in the dental pulp of 800-year-old French cats. Clin Infect Dis 39:1391–1394

Li HC, Fujiyoshi T, Lou H, Yashiki S, Sonoda S, Cartier L, Nunez L, Munoz I, Horai S, Tajima K (1999) The presence of ancient human T-cell lymphotropic virus type I provirus DNA in an Andean mummy. Nat Med 1428–1432

Lindahl T (1993) Instability and decay of the primary structure of DNA. Nature 362:709–715

Loreille O, Roumat E, Verneau O, Bouchet F, Hanni C (2001) Ancient DNA from Ascaris: extraction amplification and sequences from eggs collected in coprolites. Int J Parasitol 31:1101–1106

Marshall WF, Telford SR, Rys PN, Rutledge BJ, Mathiesen D, Malawista SE, Spielman A, Persing DH (1994) Detection of *Borrelia burgdorferi* DNA in museum specimens of *Peromyscus leucopus*. J Infect Dis 170:1027–1032

Matuschka FR, Ohlenbusch A, Eiffert H, Richter D, Spielman A (1996) Characteristics of Lyme disease spirochetes in archived European ticks. J Infect Dis 174:424–426

Mays S, Taylor M (2003) A first prehistoric case of tuberculosis from Britain. Int J Osteoarchaeol 2003:189–196

Mays S, Fysh E, Taylor GM (2002) Investigation of the link between visceral surface rib lesions and tuberculosis in a Medieval skeletal series from England using ancient DNA. Am J Phys Anthropol 119:27–36

Montiel R, Garcia C, Canadas MP, Isidro A, Guijo JM, Malgosa A (2003) DNA sequences of *Mycobacterium leprae* recovered from ancient bones. FEMS Microbiol Lett 226:413–414

Nickle DC, Learn GH, Rain MW, Mullins JI, Mittler JE (2002) Curiously modern DNA for a "250 million-year-old" bacterium. J Mol Evol 54:134–137

Ou CY, Moore JL, Schochetman G (1991) Use of UV irradiation to reduce false positivity in polymerase chain reaction. Biotechniques 10:442, 444, 446

Pääbo S (1985) Molecular cloning of Ancient Egyptian mummy DNA. Nature 314:644–645

Papagrigorakis MJ, Yapijakis C, Synodinos PN, Baziotopoulou-Valavani E (2006) DNA examination of ancient dental pulp incriminates typhoid fever as a probable cause of the Plague of Athens. Int J Infect Dis 10:206–214

Poinar HN, Hoss M, Bada JL, Pääbo S (1996) Amino acid racemization and the preservation of ancient DNA. 272:864–866

Poinar HN, Schwarz C, Qi J, Shapiro B, Macphee RD, Buigues B, Tikhonov A, Huson DH, Tomsho LP, Auch A, Rampp M, Miller W, Schuster SC (2006) Metagenomics to paleogenomics: large-scale sequencing of mammoth DNA. Science 311:392–394

Pusch CM, Giddings I, Scholz M (1998) Repair of degraded duplex DNA from prehistoric samples using *Escherichia coli* DNA polymerase I and T4 DNA ligase. Nucleic Acids Res 26:857–859

Raoult D, Aboudharam G, Crubezy E, Larrouy G, Ludes B, Drancourt M (2000) Molecular identification by "suicide PCR" of *Yersinia pestis* as the agent of Medieval Black Death. Proc Natl Acad Sci USA 97:12800–12803

Raoult D, Dutour O, Houhamdi L, Jankauskas R, Fournier PE, Ardagna Y, Drancourt M, Signoli M, La VD, Macia Y, Aboudharam G (2006) Evidence for louse-transmitted diseases in soldiers of Napoleon's Grand Army in Vilnius. J Infect Dis 193:112-120

Reid AH, Fanning TG, Hultin JV, Taubenberger JK (1999) Origin and evolution of the 1918 "Spanish" influenza virus hemagglutinin gene. Proc Natl Acad Sci USA 96:1651–1656

Reid AH, Fanning TG, Janczewski TA, Taubenberger JK (2000) Characterization of the 1918 "Spanish" influenza virus neuraminidase gene. Proc Natl Acad Sci USA 97:6785–6790

Rhodes AN, Urbance JW, Youga H, Corlew-Newman H, Reddy CA, Klug MJ, Tiedje JM, Fisher DC (1998) Identification of bacterial isolates obtained from intestinal contents associated with 12,000-year-old mastodon remains. Appl Environ Microbiol 64:651–658

Rohland N, Hofreiter M (2007) Comparison and optimization of ancient DNA extraction. Biotechniques 42:343–352

Rollo F, Luciani S, Canapa A, Marota I (2000) Analysis of bacterial DNA in skin and muscle of the Tyrolean iceman offers new insight into the mummification process. Am J Phys Anthropol 111:211–219

Rothschild BM, Martin LD, Lev G, Bercovier H, Bar-Gal GK, Greenblatt C, Donoghue H, Spigelman M, Brittain D (2001) Mycobacterium tuberculosis complex DNA from an extinct bison dated 17,000 years before the present. Clin Infect Dis 33:305–311

Salo WL, Aufderheide AC, Buikstra J, Holcomb TA (1994) Identification of *Mycobacterium tuberculosis* DNA in a pre-Columbian Peruvian mummy. Proc Natl Acad Sci USA 91:2091–2094

Serre D, Hofreiter M, Pääbo S (2004) Mutations induced by ancient DNA extracts? Mol Biol Evol 21:1463–1467

Spencer M, Howe CJ (2004) Authenticity of ancient-DNA results: a statistical approach. Am J Hum Genet 75:240–250

Spigelman M, Donoghue HD (2001) Brief communication: unusual pathological condition in the lower extremities of a skeleton from ancient Israel. Am J Phys Anthropol 114:92–93

Spigelman M, Lemma E (1993) The use of the polymerase chain reaction to detect *Mycobacterium tuberculosis* in ancient skeletons. Int J Osteoarchaol 3:143

Stead WW, Eisenach KD, Cave MD, Beggs ML, Templeton GL, Thoen CO, Bates JH (1995) When did Mycobacterium tuberculosis infection first occur in the New World? An important question with public health implications. Am J Respir Crit Care Med 151:1267–1268

Taubenberger JK, Reid AH, Krafft AE, Bijwaard KE, Fanning TG (1997) Initial genetic characterization of the 1918 "Spanish" influenza virus. Science 275:1793–1796

Taubenberger JK, Reid AH, Lourens RM, Wang R, Jin G, Fanning TG (2005) Characterization of the 1918 influenza virus polymerase genes. Nature 437:889–893

Taylor GM, Crossey M, Saldanha JA, Waldron T (1996) Detection of *Mycobacterium tuberculosis* bacterial DNA in medieval human skeletal remains using polymerase chain reaction. J Archaeol Sci 1996:789–798

Taylor GM, Rutland P, Molleson T (1997) A sensitive polymerase chain reaction method for the detection of *Plasmodium* species DNA in ancient human remains. Ancient Biomolecules 1:193–203

Taylor GM, Goyal M, Legge AJ, Shaw RJ, Young D (1999) Genotypic analysis of *Mycobacterium tuberculosis* from medieval human remains. Microbiology 145:899–904

Vreeland RH, Rosenzweig WD, Powers DW (2000) Isolation of a 250 million-year-old halotolerant bacterium from a primary salt crystal. Nature 407:897–900

Wier A, Dolan M, Grimaldi D, Guerrero R, Wagensberg J, Margulis L (2002) Spirochete and protist symbionts of a termite (*Mastotermes electrodominicus*) in Miocene amber. Proc Natl Acad Sci USA 1410–1413

Willerslev E, Cooper A (2005) Ancient DNA. Proc R Soc London B Biol Sci 272:3–16

Zink A, Reischl U, Wolf H, Nerlich AG (2000) Molecular evidence of bacteremia by gastrointestinal pathogenic bacteria in an infant mummy from ancient Egypt. Arch Pathol Lab Med 124:1614–1618

Zink AR, Reischl U, Wolf H, Nerlich AG (2002) Molecular analysis of ancient microbial infections. FEMS Microbiol Lett 213:141–147

Zink AR, Sola C, Reischl U, Grabner W, Rastogi N, Wolf H, Nerlich AG (2003) Characterization of Mycobacterium tuberculosis complex DNAs from Egyptian mummies by spoligotyping. J Clin Microbiol 41:359–367

Zink AR, Spigelman M, Schraut B, Greenblatt CL, Nerlich AG, Donoghue HD (2006) Leishmaniasis in ancient Egypt and Upper nubia. Emerg Infect Dis 12:1616–1617

Chapter 5
Histologic Detection of Past Pathogens

Hubert Lepidi

Abstract In recent years, histologic methods have been employed for the detection of infectious past pathogens (viruses, bacteria, parasites, or fungi) in ancient tissues. The goal of these palaeopathological studies is to further our understanding of the origin and spread of infectious diseases. Microorganisms can be visualised in preserved ancient tissues after different mummification processes, or in tissue sections from old paraffin blocks. The first step is the examination of paraffin sections with routine staining. Because organisms are often difficult to see in tissue sections, several special stains have been developed to visualise them. However, these histological stains are not specific. Electron microscopy may allow the detection of very small organisms such as viruses; the accurate identification of organisms is in some cases based on a specific morphology at the ultrastructural level. Depending on tissue storage conditions, immunohistological methods such as immunohistochemistry and immunofluorescence allow specific detection of microorganisms if antigenic epitopes are well preserved.

5.1 Introduction

The first step in the diagnosis of any infectious disease from recent or ancient tissue specimens is examination of tissue sections stained with hematoxylin and eosin (H&E). This histologic examination allows recognition of specific tissue and cytopathic changes, as well as consistent patterns of inflammation, and detection of microorganisms in H&E-stained sections. However, detection of microorganisms often poses a challenge for the histopathologist. Some microorganisms are too small to be seen easily by light microscopy, while larger-sized organisms may not be clearly distinguishable on H&E-stained sections because

Hubert Lepidi
Unité des Rickettsies, CNRS UMR 6020, Faculté de Médecine, Université de la Méditerranée, 27 Boulevard Jean Moulin, 13385 Marseille Cedex 5, France
E-mail: Hubert.Lepidi@medecine.univ-mrs.fr

D. Raoult and M. Drancourt (eds.), *Paleomicrobiology: Past Human Infections.*

they are obscured by surrounding tissue elements. For these reasons, numerous special stains have been developed to detect organisms in paraffin sections. The stainability of a microorganism by a particular method does not mean that the organism can be identified accurately because many other organisms may show the same staining reaction. However, application of several judiciously chosen specific stains can allow a skilled observer to make a rapid preliminary identification of many organisms based on their morphology (Woods and Walker 1996). In practise, six special stains can be used to detect microorganisms in paraffin sections: Giemsa, Gram, periodic acid-Schiff (PAS), Grocott-Gomori methenamine silver (GMS), Warthin-Starry, and Ziehl-Neelsen stains (Lepidi et al. 2002a). The Giemsa stain is the most sensitive, allowing researchers to detect most microorganisms and study their morphology. The most popular and widely used histologic method for detection of bacteria is the tissue Gram stain. To a certain extent the Gram stain aids further identification of the organism because bacteria can be classified according to their Gram stainability as Gram-positive or Gram-negative bacteria. GMS and PAS stains detect most fungi. The Warthin-Starry stain is among the most sensitive methods for detection of microorganisms but is difficult to interpret (Lepidi et al. 2000). The Ziehl-Neelsen stain is used for detection of acid-fast bacteria, especially mycobacteria. The pattern of organisation should also be considered – for example, staphylococci and streptococci tend to gather together in clusters and in chains, respectively.

Electron microscopy can be useful in recognising and identifying microorganisms (Yardley and Hendrix 1961). Although interpretation may be somewhat hindered by suboptimal tissue preservation, examination by transmission electron microscopy may reveal the identity of a microorganism based on its specific morphology at the ultrastructural level. This morphological method has been employed in ancient tissues, e.g. in the identification of *Variola* virus in old formalin-fixed tissues and in an Italian mummy from the sixteenth century, or in identification of treponemes in a renaissance Italian mummy with syphilis (Fornaciari and Marchetti 1986; Fornaciari et al. 1989; Schoepp et al. 2004).

In the 1980s, immunohistology revolutionised histopathology, particularly with regard to the categorisation of solid tumours and haematopoietic neoplasms. Immunohistological methods are based on detection of antigenic determinants in tissue sections. These techniques mainly encompass immunohistochemistry and immunofluorescence. For several years, immunohistology has also been used for the identification of infectious agents. After histochemical staining, immunohistology is the most commonly used ancillary diagnostic technique for the detection of microorganisms in histologic sections. Moreover, only immunohistological methods provide specific detection of microorganisms. These techniques use monoclonal or polyclonal antibodies directed against specific microbial antigens. Polyclonal antibodies are produced by different cells and, as a consequence, are immunochemically dissimilar. They react with various epitopes on the antigen against which they are raised. The animals most frequently used for the production of polyclonal antibodies are rabbits and goats (Lepidi et al. 2000, 2003b, 2004). Several other animals can be used to raise polyclonal antibodies. In contrast, monoclonal antibodies are produced by clones of plasma cells. Antibodies from a given clone are immunohistochemically identical and react with a specific epitope on the antigen against which they are raised. Mice are currently used almost exclusively for the production of monoclonal antibodies. Once bound, the antibodies are detected by use of either fluorescent or chromogenic signal amplification. Immunofluorescence methods are usually performed on freshly frozen tissue, whereas immunoperoxidase methods are usually performed on formalin-fixed, paraffin-embedded tissues. These methods are useful for the detection of fastidious or noncultivatable microorganisms, or when the tissues have been fully fixed, for differentiating between morphologically similar microorganisms or cytopathic effects, and for the detection of highly infectious microorganisms involved in outbreaks of infection. Detection of fastidious microorganisms by use of ancillary methods is particularly important because they may go undetected in the microbiology laboratory. For example, *Coxiella burnetii* or *Trophyrema whipplei*, the causative agents of Q fever and Whipple's disease, respectively, are usually not cultured, but they can be readily detected in tissue samples from infected patients by use of immunoperoxidase methods (Lepidi et al. 2002b, 2003a, 2003b, 2004). The specificity imparted by immunohistological stains has been used to differentiate morphologically similar microorganisms such as *Rickettsia conorii* and *Rickettsia africae* (Lepidi et al. 2006). Similarly, these immunohistochemical methods have been used to differentiate morphologically cytopathic effects, such as those produced by *T. whipplei* and *Mycobacterium avium* or *Mycobacterium intracellulare* (Lepidi et al. 2003a). Immunohistology may also be more sensitive for detection of microorganisms that are difficult to locate in histologic sections (Toulaymat et al. 1999).

A potential pitfall of immunohistological methods is the failure to detect microorganism antigen because of prolonged storage of the tissue in fixatives such as formaldehyde. In these cases, additional steps for antigen retrieval must be performed, such as protease digestion, or heating in a microwave oven or in a 95–99°C water bath with sodium citrate buffer. However, specific antibodies for immunohistological staining are commercially available for a few bacteria. Polyclonal mouse or rabbit antibodies against microorganisms can be generated in laboratories if microorganisms are cultivated (Lepidi et al. 2000, 2003b, 2004, 2006).

Immunohistochemistry and immunofluorescence have been successfully employed in a number of palaeopathological studies, demonstrating that antigenic properties in ancient tissues can be preserved, and that preservation is related to different mummification processes and body storage conditions (Bruschi et al. 2006; Ciranni et al. 1999; Fornaciari and Marchetti 1986; Fornaciari et al. 1989). Immunohistochemistry can also be applied to old paraffin blocks in which antigenic epitopes are well preserved. This method has been employed to detect bacteria such as *T. whipplei* and *Rickettsia rickettsii* in one-century-old paraffin blocks from autopsy cases, or viruses and parasites in mummies (Bruschi et al. 2006; Dumler 1991; Dumler et al. 2003; Fornaciari and Marchetti 1986). However, although immunohistological methods seem an attractive option with which to detect and visualise microorganisms in ancient tissues, antigenic determinants in such tissues are often impaired or destroyed. This important technical limitation probably explains the very few studies concerning the detection of past pathogens in the literature.

References

Bruschi F, Masetti M, Locci MT, Ciranni R, Fornaciari G (2006) Cysticercosis in an Egyptian mummy of the late Ptolemaic period. Am J Trop Med Hyg 74:598–599

Ciranni R, Castagna M, Fornaciari G (1999) Goiter in an eighteenth-century Sicilian mummy. Am J Phys Anthropol 108:427–432

Dumler JS (1991) Fatal Rocky Mountain spotted fever in Maryland – 1901. JAMA 265:718

Dumler JS, Baisden BL, Yardley JH, Raoult D (2003) Immunodetection of *Tropheryma whipplei* in intestinal tissues from Dr. Whipple's 1907 patient. N Engl J Med 348:1411–1412

Fornaciari G, Marchetti A (1986) Intact smallpox virus particles in an Italian mummy of sixteenth century. Lancet 8507:625

Fornaciari G, Castagna M, Tognetti A, Tornaboni D, Bruno J (1989) Syphilis in a renaissance Italian mummy. Lancet 8663:614

Lepidi H, Fournier PE, Raoult D (2000) Quantitative analysis of valvular lesions during *Bartonella endocarditis*. Am J Clin Pathol 114:880–889

Lepidi H, Durack DT, Raoult D (2002a) Diagnostic methods: current best practices and guidelines for histologic evaluation in infective endocarditis. Infect Dis N Am 16:339–361

Lepidi H, Costedoat N, Piette JC, Harlé JR, Raoult D (2002b) Immunohistological detection of *Trophyrema whipplei* (Whipple bacillus) in lymph nodes. Am J Med 113:334–336

Lepidi H, Fenollar F, Gerolami R, Mege JL, Bonzi MF, Chappuis M, Sahel J, Raoult D (2003a) Whipple's disease: immunospecific and quantitative immunohistochemical study of intestinal biopsies. Hum Pathol 34:589–596

Lepidi H, Houpikian P, Liang Z, Raoult D (2003b) Cardiac valves in patients with Q fever endocarditis: microbiological, molecular and histologic studies. J Infect Dis 187:1097–1106

Lepidi H, Fenollar F, Dumler JS, Gauduchon V, Chalabreysse L, Bammert A, Bonzi MF, Thivolet-Béjui, Vandenesch F, Raoult D (2004) Cardiac valves in patients with Whipple endocarditis: microbiological, molecular, quantitative histologic, and immunohistochemical studies of 5 patients. J Infect Dis 190:935–945

Lepidi H, Fournier PE, Raoult D (2006) Histologic features and immunodetection of African tick-bite fever eschar. Emerg Infect Dis 12:1332–1337

Schoepp RJ, Morin MD, Martinez MJ, Kulesh DA, Hensley L, Geisbert TW, Brady DR, Jahrling PB (2004) Detection and identification of *Variola* virus in fixed human tissue after prolonged archival storage. Lab Invest 84:41–48

Toulaymat M, Marconi S, Garb J, Otis C, Nash S (1999) Endoscopic biopsy pathology of *Helicobacter pylori* gastritis. Comparison of bacterial detection by immunohistochemistry and Genta stain. Arch Pathol Lab Med 123:778–781

Woods GL, Walker DH (1996) Detection of infection or infectious agents by use of cytologic and histologic stains. Clin Microbiol Rev 9:382–404

Yardley J, Hendrix TR (1961) Combined electron and light microscopy in Whipple's disease: demonstration of "bacillary bones" in the intestine. Bull Johns Hopkins Hosp 109:80–98

Part II
Ancient Microorganisms

Bacteria

Chapter 6
Palaeomicrobiology of Tuberculosis

Helen D. Donoghue

Abstract The study of tuberculosis palaeomicrobiology has proved to be most rewarding. Due to the characteristic palaeopathological lesions, tuberculosis was recognised in archaeological material and was the first infectious disease to be studied by modern biomolecular methods. The combination of a tough bacterial cell wall and GC-rich DNA has resulted in excellent DNA preservation in some specimens. A wide range of specific molecular diagnostic and typing methods, developed by clinical microbiologists, are available. These have been applied successfully to archaeological material, resulting in the genotyping of the infecting organisms. There has been a fruitful interaction with modern genomic studies, and ancient findings support current views on the evolution of the species in the *Mycobacterium tuberculosis* complex. Questions remain to be answered, including the nature of pre-Columbian tuberculosis in the Americas, and the evolution of tuberculosis in animals. The important topics of interactions with other pathogenic microbes, and the host, are now being explored.

6.1 Introduction

Tuberculosis is a major scourge in the world today and the World Health Organisation estimates that around 2 billion people, about one-third of the world's total population, are infected with tubercle bacilli (WHO 2006). Only about 10% of infected persons will become ill with active disease, and those with weakened immune systems such as the very young and old, people who suffer from malnourishment, other diseases, physical or mental stress, or other immunosuppressive conditions, are more likely to suffer from the disease. The extremely high level of latent infection is an indication of long-term co-existence of human host and bacterial

Helen D. Donoghue
Centre for Infectious Diseases and International Health, Department of Infection, Windeyer Institute of Medical Sciences, UCL, 46 Cleveland Street, London W1T 4JF, UK
E-mail: h.donoghue@ucl.ac.uk

D. Raoult and M. Drancourt (eds.), *Paleomicrobiology: Past Human Infections.*

pathogen (Hirsh et al. 2004), and the evolution of the causative organism is an active area of research.

6.1.1 Causative Organisms

Tuberculosis is caused by a group of closely related bacterial species termed the *Mycobacterium tuberculosis* (MTB) complex. Other mycobacterial species are widespread in the environment but members of the MTB complex are obligate pathogens. All mycobacteria have DNA that is rich in guanine and cytosine, and all have a lipid-rich cell wall that is hydrophobic and extremely resistant to damage and degradation. Many of the mycobacteria, including the MTB complex, are very slow growing. The pathogenic species are able to survive and grow within macrophages, which enables them to evade the host immune system. An active cell-mediated immune response is required to contain and kill the tubercle bacilli.

The principal cause of human tuberculosis is *M. tuberculosis. Mycobacterium bovis* has a wider host range and is the main cause of tuberculosis in other animal species. Humans can become infected by *M. bovis*, usually via milk, milk products or meat from an infected animal. Other members of the MTB complex include the human pathogens *Mycobacterium canettii*, *Mycobacterium africanum*, and several species associated with particular animals such as voles (*Mycobacterium microti*), goats and deer (*Mycobacterium caprae*), and seals (*Mycobacterium pinnipedii*).

6.1.2 Natural Course of Tuberculosis

Tuberculosis (TB) is spread principally by infectious aerosols, which are released from the lungs of an infected person who has pulmonary disease. In the alveolus of the lung, inhaled tubercle bacilli are ingested by macrophages and are normally contained by the host immune response. This leads to granuloma formation and eventually to calcified lesions. Spread of the bacteria within a year of initial infection results in primary disease. During the lifetime of the host the organisms may remain dormant but viable for decades and, if the immune response is compromised, the bacteria may escape into the lungs causing re-activated pulmonary tuberculosis. In a minority of cases the bacteria spread to other host tissues via the lymphatic system and blood, thereby becoming disseminated throughout the body, causing extra pulmonary tuberculosis, e.g. miliary TB or meningitis (Garay 1996).

Occasionally tuberculosis can be acquired by ingestion of infected animal products, causing intestinal tuberculosis. This can also result from swallowing infected sputum. Therefore both urine and faeces may contain tubercle bacilli and act as vehicles of infection. Extra pulmonary tuberculosis can result in infected lymph glands. Cervical lymphadenitis and skin lesions were previously known as scrofula or lupus vulgaris, and were often associated with *M. bovis* infections (Grange 1995). In developing countries with little or no effective prevention or treatment, it is estimated that around 6% of human tuberculosis cases are caused by *M. bovis*, with a higher proportion in children (Hardie and Watson 1992; O'Reilly and Daborn 1995).

6.1.3 Laboratory Diagnosis of Tuberculosis

The basic method of diagnosis of tuberculosis in patients with lung disease is to microscopically examine stained smears of the sputum and, if facilities are available, to isolate and grow a culture of the infecting organism. Because of the hydrophobic bacterial cell wall, the organisms can be stained only by special procedures, such as heating the slide, that enable the stain to penetrate into the cell. Subsequent decoloration by cold acid and/or alcohol will fail to remove the stain, so this group of bacteria is described as acid-fast. However, this method is not sufficiently sensitive to detect low levels of bacteria, or to diagnose cases of extra pulmonary disease. Culture is still the universally accepted method of confirmation of clinical diagnosis, but requires specialised facilities to protect laboratory staff. In addition, the organisms take several weeks to grow. Therefore, members of the MTB complex were amongst the first microorganisms for which molecular diagnostic methods were devised. The complete genome sequences are available for several mycobacterial species and this has increased the number of methods available (Drobniewski et al. 2003).

6.1.4 Molecular Diagnosis of Tuberculosis

Organisms in the MTB complex have many repetitive sequences in their DNA. These are of no known function and it is probable that the majority are not transcribed. There are several insertion sequences, including IS*6110* and IS*1081*, normally present in multiple copies within the cell (Gordon et al. 1999). Following the development of the polymerase chain reaction (PCR), assays targeted regions of these insertion elements specific to the MTB complex were devised (Eisenach et al. 1990). Care is needed as several environmental species have similar sequences, e.g. in the 16S ribosomal DNA locus, insertion sequences and other widespread alleles such as the 65-kDa heat shock protein gene, *hsp65* (Dziadek et al. 2001; Huggett et al. 2003). It was realised that there were three principal genetic groups within the MTB complex, based upon two functionally neutral single nucleotide polymorphisms (SNPs) in the catalase-peroxidase-encoding gene *katG* and a subunit of the DNA gyrase gene *gyrA* (Sreevatsan et al. 1997; Mathema et al. 2006). Whole genome sequencing has led to a revision of our earlier ideas about the evolution of the *M. tuberculosis* complex (Brosch et al. 2002) and it appears that this group has accumulated deletions over time that can be used to distinguish individual species (Parsons et al. 2002). It is also now clear that the human pathogen *M. tuberculosis* has evolved from a more ancestral lineage than *M. bovis*.

6.1.5 Molecular Typing Methods

Different strains of *M. tuberculosis* may have up to 26 copies of the IS*6110* element, or rarely, none at all. *M. bovis* has a single copy of IS*6110*. Both species have six copies of IS*1081* (Collins and Stephens 1991). An early method of molecular fingerprinting was to target the IS*6110* locus by restriction enzyme digestion, separate the fragments on a gel and to analyse the restriction fragment length polymorphism (RFLP). However, this expensive and time-consuming method is being replaced by PCR-based techniques (van Soolingen 2001). Spoligotyping is based on the direct repeat (DR) region of the MTB complex organisms (Kamerbeek et al. 1997). PCR primers in the DR region amplify up to 43 unique spacer regions that lie between each DR locus. The presence of these individual spacers is visualised by means of dot-blot hybridisation on a membrane, giving a fingerprint. The more ancestral lineages have a complete spoligotype, but different strains commonly show deletions. The loss of spacers is unidirectional, so the data can be used for evolutionary studies. Spoligotyping clearly distinguishes *M. bovis* from *M. tuberculosis*, and different families of strains are defined by characteristic patterns. An international database is available at www.pasteur-guadeloupe.fr/tb/spoldb4 (Brudey et al. 2006). Further typing is possible based upon variable number tandem repeat (VNTR) loci, and mycobacterial interspersed repetitive units (MIRU), which are tandem repeats of 40–100 bp located in microsatellite regions around the chromosome (Barnes and Cave 2003). A combination of spoligotyping and MIRU typing has the greatest discriminatory power. Meanwhile, research continues on typing based on genomic deletion analysis and SNPs (Baker et al. 2004; Mathema et al. 2006).

6.2 Tuberculosis in the Past

6.2.1 Palaeopathology of Tuberculosis

The disease has long been recognised by the characteristic changes that occur in the spine, i.e. gibbus formation leading to Pott's disease. In addition, periosteal reactive lesions on tubular bones and osteomyelitis are indications of tuberculosis (Ortner and Putschar 1981). Such palaeopathological changes have been reported in pre-dynastic

(3500–2650 B.C.) Egypt (Zink et al. 2001); middle Neolithic Sweden (3200–2300 B.C.) – culturally associated with the Funnel Beaker Culture, the earliest cattle breeders in Sweden (Nuorala et al. 2004); and middle Neolithic Italy at the beginning of the fourth millennium B.C. (Formicola et al. 1987; Canci et al. 1996). The disease was present in Asia, for example, in northeast Thailand at an Iron Age site dated from 2,500 to 1,700 years BP (Tayles and Buckley 2004). Tuberculosis was also present in China 2,000 years ago (Fusegawa et al. 2003). Erosive lesions suggestive of tuberculosis have been found on fossil fauna from the natural Trap Cave in Wyoming, dated from the 17,000 to 20,000 year level (Rothschild and Martin 2003). Initially, it was believed that humans acquired tuberculosis from animals, especially after domestication (Steinbock 1976; Manchester 1984; Clark et al. 1987), but now we know that human tuberculosis is more ancestral (Armelagos and Harper 2005). Animal domestication is likely to have been important in sustaining a denser human population, enabling *M. tuberculosis* to become endemic (Weiss and McMichael 2004; Armelagos et al. 2005).

6.2.2 History of Tuberculosis

Recognisable descriptions of tuberculosis are found in ancient Egyptian, Greek and Roman texts, and continued throughout history (Haas and Haas 1996; Roberts and Buikstra 2003a, 2003b). Tuberculosis reached epidemic levels in Europe during the Industrial Revolution and was responsible for one in four deaths from the sixteenth to eighteenth centuries (Hutás 1999). In recent years it has become accepted that tuberculosis existed in the Americas before European contact (Daniel 2000; Gómez I Prat and Mendonça de Souza 2003; Mackowiak et al. 2005), although debate continues over which species was responsible (see below). The disease is thought to have reached the Americas via animals (Rothschild and Martin 2006) or early nomads (Daniel 2000) who crossed the Beringa land bridge at least 10,000 years ago. However, the nature of pre-Columbian tuberculosis is still unknown (Wilbur and Buikstra 2006), and the suggestion that more virulent strains of the tubercle bacilli originated in Europe and were spread to the Far East and the Americas during the colonial expansions from the fifteenth century onwards (Clarke et al. 1987), still awaits confirmation.

6.3 Work on *Mycobacterium tuberculosis* Complex Ancient DNA

6.3.1 Recommended Good Laboratory Practice

DNA is an unstable molecule, and modern DNA sequences will always outnumber those of ancient DNA (aDNA) in any sample. Therefore, stringent precautions must be taken to reduce extraneous contamination to a minimum. These should be applied during the initial removal of samples from the archaeological site (Spigelman and Greenblatt 1998), and throughout all subsequent examinations. Criteria for mammalian aDNA work have been devised for use in the verification of findings (O'Rourke et al. 2000; Hofreiter et al. 2001; Pääbo et al. 2004). Due to the tendency of aDNA to fragment, there should be an inverse correlation between length of target sequence and amplification efficiency, with claims of long amplicons scrutinised. Results should be repeated in a second extract, and verified in an independent laboratory. It is also recommended that the number of amplifiable DNA molecules be quantified, and that PCR products be cloned and sequenced. For palaeomicrobiological MTB complex DNA studies, modified criteria may be more appropriate (Table 6.1).

Table 6.1 Procedures used to determine authenticity of pathogenic mycobacterial ancient DNA (MTB complex aDNA). PCR polymerase chain reaction, SNP single nucleotide polymorphism

Procedure	Comments and examples
The choice of sampling sites should be determined by the natural history of the disease	For pulmonary tuberculosis select inner rib surfaces, or lung tissue; for disseminated tuberculosis sample vascularised tissues
The size and copy number of the MTB aDNA alleles sought should be appropriate for the sampling site	Sites that give protection from the natural decay process are the dental pulp region and the ends of the long bones; abdominal tissue is likely to pose the greatest challenge
Pre- and post-PCR activities must be strictly separated	To prevent cross-contamination
Ensure no modern DNA is used in aDNA laboratory	Modern DNA is non-fragmented and if present will inevitably be the major PCR amplicon
Use multiple negative controls of DNA extraction and PCR	This detects cross-contamination of reagents and by glove-tip
Confirm results by replication within laboratory	Results are often inconsistent due to uneven distribution of pathogen. Replication should be tested with a repeat DNA extract if possible
Independent confirmation of results by external laboratory	Many replicates are advised as discrepant results may be genuine
Results should be consistent with natural history of the infectious disease	TB likely to be disseminated in infants, and pulmonary in adults
An inverse relationship between fragment size and quantity of PCR amplicon should be observed	GC-rich microbial templates such as MTB can yield remarkably large PCR amplicons e.g. ~300 bp if preservation is good
aDNA sequence data should make phylogenetic sense	Direct sequencing is adequate for GC-rich mycobacterial aDNA
Cloning or multiple sequencing	This may be useful when investigating SNPs and MTB genetic variations

6.3.2 DNA Extraction

The first stage is to disaggregate the sample. Published protocols include drilling bone, grinding in a pestle and mortar, successive freezing in liquid nitrogen and thawing, demineralisation with proteinase K and EDTA, and incubation in a lysis buffer based on a guanidium salt. Pre-incubation with the reagent *N*-phenacylthiazolium bromide (PTB) enables DNA to be released from material with glucose-derived protein cross-links that can form over time (Poinar et al. 1998). After release from the specimen, DNA is normally captured onto silica (Boom et al. 1990; Höss and Pääbo 1993), and repeated silica extraction is a simple way to remove inhibitors (Kemp et al. 2006). Alternatively, DNA is precipitated by isopropanol, which also removes inhibitors (Hänni et al. 1995). Finally, DNA is eluted or re-hydrated into solution. DNA extracts are not stable so are often aliquoted into 'no stick' plastic tubes, before storing at −20°C or, preferably, −80°C to avoid unnecessary freezing and thawing.

6.3.3 DNA Amplification

Stringent precautions against cross-contamination must be taken, with physical and temporal separation for different stages of the process, e.g. extraction, PCR set-up and product analysis. Separate sets of pipettes should be used for PCR set-up and product analysis, and cleaned thoroughly before use. Filter tips are routinely used and all surfaces and equipment in contact with sample tubes (centrifuges, rotors, mixers, etc.) cleaned before each assay. Multiple sample blanks should be used for negative controls during the DNA extraction and water blanks included in PCR amplifications to ensure there is no contamination. PCR facilitators, such as bovine serum albumin (BSA) (Forbes and Hicks 1996) and betaine (Abu Al-Soud and Rådström 2000) are often required when amplifying aDNA. Additional *Taq* polymerase can also improve the yield (Sutlovic et al. 2005). Normally 40–45 rounds of amplification are used and further amplification in a nested PCR may be necessary.

6.4 Relationship of MTB complex aDNA to Other Markers

6.4.1 Host Proteins and DNA

In many areas of aDNA research, other markers of molecular diagenesis are used to determine the choice of specimen for examination. For example, the degree of amino acid racemisation in a specimen is taken as a measure of the extent of DNA preservation (Poinar and Stankiewicz 1999). If aDNA was found in samples with demonstrable amino acid degradation, findings were regarded sceptically. Later work suggests that both hydroxyapatite and collagen yield may be better indicators of biomolecular stability (Götherström et al. 2002).

However, the DNA of *M. tuberculosis* is intrinsically more stable than that of mammalian DNA due to its high percentage GC content. In addition, the bacterial cell wall is both persistent and protective, being lipid-rich and hydrophobic (Spigelman and Donoghue 2003; Donoghue et al. 2004), with large amounts of C60–C90 fatty acids – the so-called mycolic acids (see below) – and many unusual extractable free lipids (Minnikin et al. 2002). These provide protection from environmental extremes and also limit the permeability of the wall (Lambert 2002). This resistant cell wall is believed to be responsible for the persistence and survival of tubercle bacilli throughout the lifetime of a mammalian host and the initial decay process after death (Weed and Bagenstross 1951; Sterling et al. 2000). Therefore, MTB complex aDNA can be found in samples that otherwise appear to be poorly preserved (Donoghue and Spigelman 2006).

6.4.2 Lipid Biomarkers

All mycobacteria, including *M. tuberculosis*, have characteristic long-chain fatty acids and other cell wall components. These can be detected by high performance liquid chromatography (HPLC) and present techniques can distinguish the MTB complex from other species (Butler and Guthertz 2001). HPLC has detected cell-wall mycolic acids specific for the *M. tuberculosis* complex from archaeological specimens (Gernaey et al. 1998, 2001), and confirmed findings of *M. tuberculosis* complex DNA (Donoghue et al. 1998). More recently, mass spectroscopy (MS) has been used to detect such molecules (Mark et al. 2006). Unfortunately, the equipment and expertise are not yet widely available, yet this is potentially a method of great promise. The advantage of using HPLC or MS in ancient tuberculosis studies is that these lipid biomarkers are very stable and the methods exquisitely sensitive so the molecules can be detected directly without amplification.

6.5 The Start of Palaeomicrobiology

6.5.1 Early Molecular Studies

Tuberculosis was the first ancient infectious disease to be detected via the DNA of the causative organism. This was due to the combination of clear skeletal markers of the disease, coupled with the availability of specific molecular diagnostic methods based on PCR. The first study, on skeletal samples (Spigelman and Lemma 1993), used MTB complex-specific PCR primers that targeted a short region of

123 bp in the repetitive locus IS*6110* (Eisenach et al. 1990). It is important to use a short target sequence, preferably <130 bp, because aDNA fragments, and damage accrues over time (Pääbo et al. 2004). A repetitive sequence enhances the chance of obtaining a positive result. This first study demonstrated tuberculosis in 4 of 11 specimens that had been morphologically diagnosed with tuberculosis, including 1 from Borneo dated prior to known European contact. These findings were later repeated in independent laboratories for verification, and confirmed by sequencing (Spigelman et al. 2002). Shortly after the initial report, Salo et al. (1994) used the same primers to clone and sequence amplicons from lung tissue that had been sampled from a Peruvian mummy 1,000 years BP. This and subsequent studies (Arriaza et al. 1995; Braun et al. 1998; Konomi et al. 2002) showed that tuberculosis was undoubtedly present in the Americas before Columbus.

Salo et al. (1994) is the earliest molecular study that made use of nested PCR based on IS*6110*, giving a 97 bp amplicon. The technique was also used in the study of Mediaeval remains from a London cemetery, where the burial conditions were not ideal for DNA preservation (Taylor et al. 1996). Here, a smaller nested product of 92 bp was sought, and these primers have often been used since.

Some early molecular studies used primers that were not specific for the MTB complex, or had too large a target to give reliable results. For example, Nerlich et al. (1997) examined ancient Egyptian tissue from the New Kingdom (1,550–1,080 B.C.). PCR was carried out using primers for a 133 bp sequence from the *hsp65* gene and confirmed by sequencing. Although the primers were based on the *M. tuberculosis* sequence, the DNA from many environmental mycobacteria will amplify. The 65 kDa heat shock protein gene was used as an additional target by Crubézy et al. (1998) in their examination of skeletal remains of a pre-dynastic Egyptian (5,400 years BP) with Pott's disease. Haas et al. (2000) also used primers for *hsp65* but concluded that the IS*6110* locus had greater specificity and should be used in preference. Amplification with non-specific primers followed by sequencing of the amplicons is a strategy that has been used where the causative organisms are unknown, such as the case of the Tyrolean Iceman (Cano et al. 2000). However, there is a danger in such studies, using primers for loci such as 16s rRNA, that chimaeric amplicons will be obtained that bear no relation to any original sequence.

The 10-year anniversary of palaeomicrobiology in 2003 resulted in several reviews, which summarise the early studies (Zink et al. 2002; Donoghue et al. 2004; Drancourt and Raoult 2005).

6.5.2 Relationship of Bony Lesions to MTB complex aDNA

It is estimated that around 40% of skeletal tuberculosis cases result in tuberculosis of the spine (Aufderheide and Rodriguez Martin 1998). However, it is important to appreciate that tuberculosis of the bone is comparatively rare, and possibly occurs in only 3–5% of cases allowed to run their natural course (Resnick and Niwayama 1995). Therefore, according to the natural history of tuberculosis, in the great majority of cases there should be no skeletal lesions, and the incidence of tuberculosis was undoubtedly far higher than that suggested by the level of bony lesions observed by palaeopathologists.

Non-microbiologists, who did not appreciate the comparative rarity of skeletal tuberculosis, initially viewed early reports that bones without lesions were found to be positive for MTB complex DNA with scepticism (Baron et al. 1996; Faerman et al. 1997). As work has continued in this field, however, there are an accumulating number of studies that have reported MTB complex DNA in skeletons without pathological changes (Haas et al. 2000; Zink et al. 2001, 2003a; Mays et al. 2002; Fusegawa et al. 2003), although at a lower frequency than skeletons with lesions. There is continuing interest in the relationship of disease to the presence of bony lesions (Zink et al. 2005a; Raff et al. 2006; Santos and Roberts 2006).

It should be remembered that MTB complex DNA will be unevenly distributed within a host, and that it will always be in the minority, compared with residual human DNA and that of the commensal and saprophytic microflora associated with the skeletal remains. Therefore, a bone with typical tuberculosis pathology provides an excellent marker for a site where MTB complex DNA may be localised. Another excellent site is the pleural surface of the ribs, as these often contain MTB complex DNA due to contact with infected lungs. To detect MTB complex DNA that was present in the bloodstream, the heads of the long bones and centre of all tubular bones should be sampled, as these will contain residues of the bone marrow. The residual material in the dental pulp cavity is another excellent source of microbial DNA in disseminated infections.

6.5.3 MTB complex DNA in Populations

The examination of tuberculosis within past populations is especially worthwhile, as data are obtained from infections that were allowed to run their natural course, in the absence of effective treatment. This offers the potential to investigate the host–pathogen interaction at a molecular level. The earliest populations studied were those of ancient Egypt (Zink et al. 2001, 2003a), and it is clear that tuberculosis infections were relatively frequent, from predynastic (ca. 3,500–2,650 B.C.) to the Late Period (ca. 1,450–500 B.C.). In some cases the age and sex of the individuals could be determined, but the main conclusion drawn was that MTB complex could be detected in bones with typical palaeopathology, non-specific palaeopathology, and without visible palaeopathology. The tombs examined were mainly the ‘Tombs of the Nobles’ in Thebes-West, and the relatively high incidence of disease was related to the dense crowding in the city at a time of prosperity.

An early study of pre-Columbian tuberculosis in Northern Chile examined 483 skeletons, dating from 2,000 B.C. to 1,500 A.D. (Arriaza et al. 1995). Morphological evidence of tuberculosis was found mainly in the period 500–1,000 A.D., which correlated with fully agaropastoral societies, and about 2% showed tuberculosis lesions. However, molecular data were obtained from only one 12-year-old girl with Pott’s disease.

The discovery of a crypt containing 263 wholly or partially naturally mummified bodies in the Hungarian town of Vác, led to an on-going study of tuberculosis at a time when the disease was reaching epidemic levels just before the industrial revolution in that part of Europe (Pap et al. 1999). As there is a contemporaneous archive of the individuals buried in the crypt, it is possible to determine the age, sex, name, family relationship and even the occupation in some cases. In some individuals the preservation of the human remains was remarkable (Fig. 6.1a,b). It was soon discovered that tuberculosis was widespread

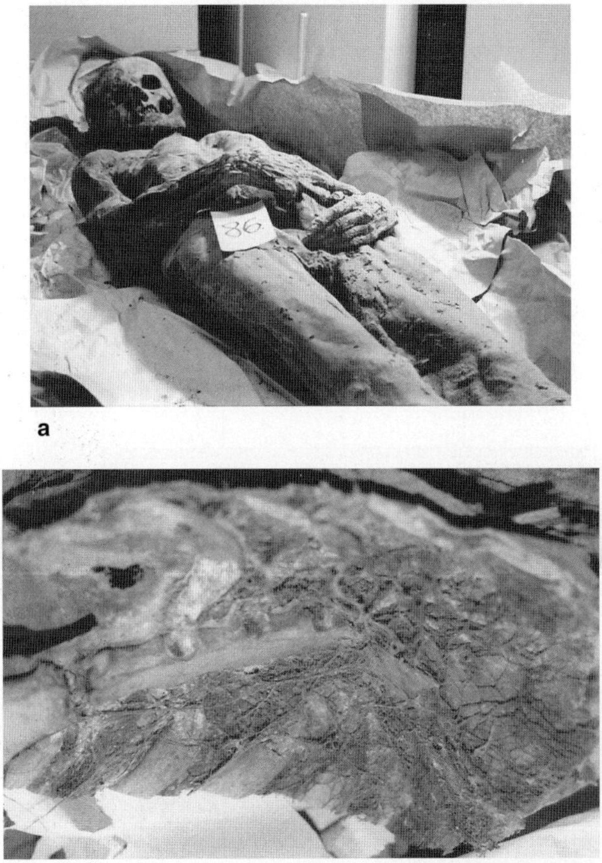

a

b

Fig. 6.1a,b Naturally-mummified individuals from eighteenth century Vác, Hungary. **a** Whole body of a 76-year-old man who died in 1796. Tissue from his right chest contained *Mycobacterium tuberculosis* ancient DNA (*M. tuberculosis* aDNA). Other samples were negative. **b** Inner surface of chest cavity of partially mummified 36-year-old man who died suddenly in 1808 vomiting blood, and after long-lasting spitting of blood. Note the well-preserved vascular tissue. His chest and abdomen tissues were strongly positive for *M. tuberculosis* aDNA and a radiograph showed small calcified lesions in his thorax, typical of pulmonary tuberculosis

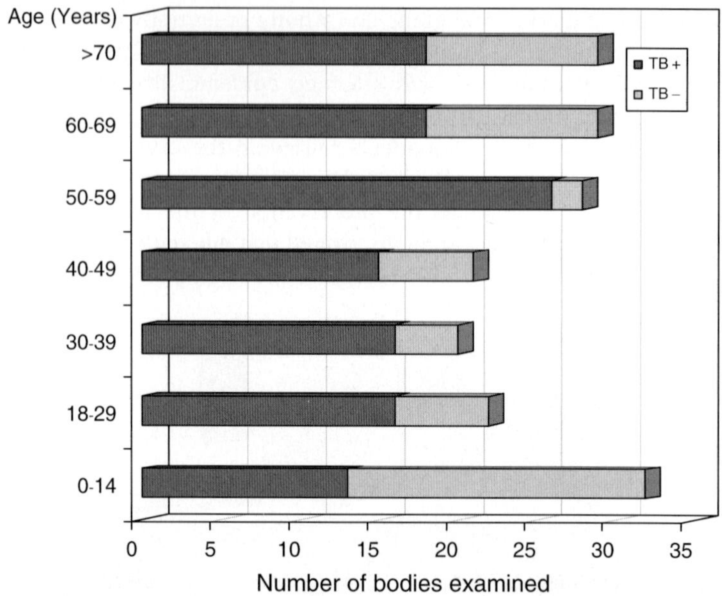

Fig. 6.2 Distribution of *M. tuberculosis* aDNA by age at death in eighteenth century mummies, in Vác, Hungary. There is a very high level of infection in all age groups, particularly in those aged 50–59 years. However, some infected individuals lived to a great age

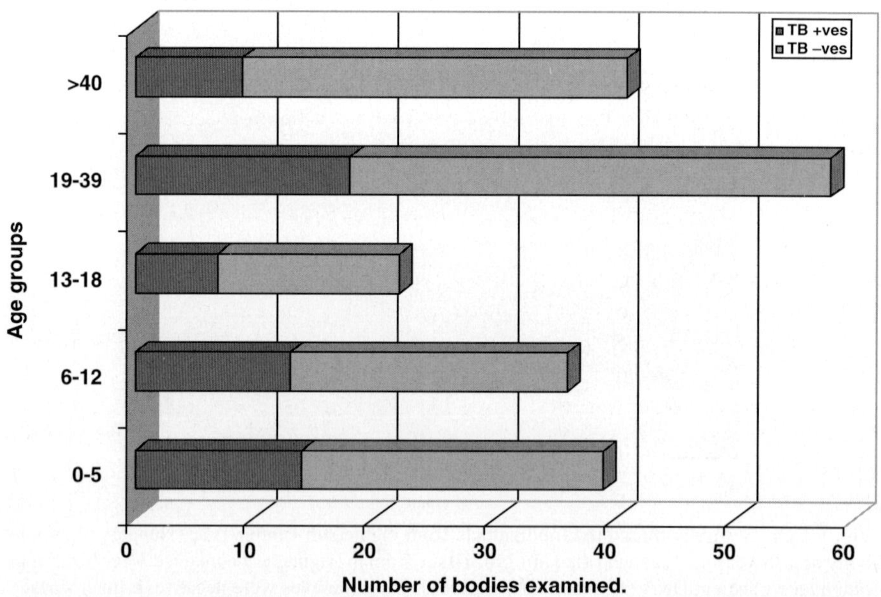

Fig. 6.3 Distribution of MTB complex DNA by age at death in Kulubnarti, early Christian Nubia. Individuals show a lower level of infection compared with eighteenth century Hungary, with highest levels in adults aged 19–39 years, and young children

in the population (Fletcher et al. 2003a), with almost every individual in some age groups infected (Fig. 6.2). Because of the availability of mummified lung tissue, it was possible to distinguish between individuals who probably died from the disease, from those who were infected but lived to a great age. Examination of tuberculosis infection with year of birth indicated that the highest rates of infection occurred during a period of planned expansion of the town with net immigration.

Another large on-going study is based on several hundred early Christian partially mummified remains from two sites in Kulubnarti, Nubia (van Gerven et al. 1981). These have been thoroughly studied by anthropologists, so there is excellent information on their age, sex and nutritional status. Tuberculosis was widespread although at a lower level than in eighteenth century Hungary (Fig. 6.3). Individuals died at a much younger age and there appeared to be more tuberculosis infection in young adults and children under 5 years of age, but the data have not yet been fully published (Spigelman et al. 2005).

6.6 Molecular Characterisation of Ancient MTB complex aDNA

6.6.1 Detection of Deletions and Spoligotypes

In an effort to characterise the strains of MTB detected by PCR, many genetic loci have been examined in addition to IS*6110*, although none are as sensitive. It is inevitable that single copy genetic loci will give less consistent results, as they will be more prone to the impact of poor DNA preservation. However, Taylor et al. (1999) attempted to determine whether *M. tuberculosis* or *M. bovis* was present in three mediaeval skeletons from the London Royal Mint site. They used primers for mtp40, a region found in most *M. tuberculosis* isolates, and obtained positive results. In addition, a locus in the *oxyR* pseudogene was amplified and sequenced, revealing a guanine residue at position 285, typical of *M. tuberculosis* but not of *M. bovis* (Sreevatsan et al. 1996). Finally, spoligotyping was performed, and showed a pattern consistent with *M. tuberculosis*. Subsequent work on material from a deserted Mediaeval village site in Yorkshire (UK) included PCR based on regions known to be deleted (DR regions) in different members of the MTB complex. In addition, genetic loci related to virulence, such as *rpoB* and *pncA*, associated with susceptibility or resistance to rifampicin and pyrazinamide, respectively, were explored for their use in characterising MTB aDNA (Mays et al. 2001; Taylor et al. 2001). This biomolecular approach, expanded by the inclusion of flanking and internal primers for the TbD1 deletion (Brosch et al. 2002), successfully identified the organisms isolated by Robert Koch and stored in a museum display case, to be of a 'modern' evolutionary lineage of *M. tuberculosis* (Taylor et al. 2003).

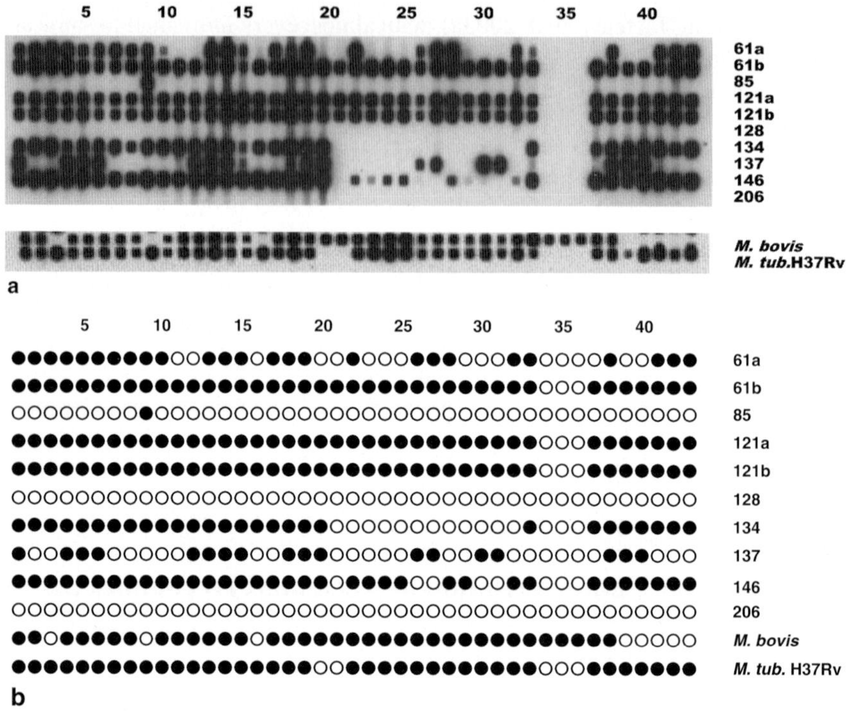

Fig. 6.4 Spoligotypes of MTB complex aDNA from different individuals in Vác, Hungary, in comparison with *M. tuberculosis* H37Rv and *Mycobacterium bovis*. The data are consistent with all individuals being infected with *M. tuberculosis* of 'modern' lineage

This biomolecular approach was adopted by Fletcher et al. (2003a, 2003b) in their study of the Vác mummies (Fig. 6.4). Both genotyping and spoligotyping were carried out, and the infecting organisms were shown to be 'modern' strains of *M. tuberculosis,* of principal genomic groups 2 and 3. Interestingly, three members of a family group apparently were infected with different strains. This study also clearly demonstrated the inverse relationship between the number of amplified copies per microlitre and increasing PCR target size, caused by fragmentation of aDNA. The nature of the tuberculosis in 85 Egyptian mummies was also investigated by the use of additional PCR target regions and spoligotyping (Zink et al. 2003b). Spoligotyping gave the most clear-cut results, and in 12 cases indicated *M. tuberculosis* or *M. africanum* patterns, but clearly not *M. bovis*.

The use of spoligotypes in MTB complex aDNA studies has initiated a discussion on whether it is appropriate to use 'consensus' spoligotypes, where positive results from repeated blots are amalgamated. As the individual spacer regions are single-copy oligonucleotides, palaeomicrobiologists tend to use consensus patterns, although this is viewed with suspicion by those responsible for maintaining the international spoligotyping database.

6.6.2 *The Search for* M. bovis *DNA in Archaeological Material*

M. bovis infects many host species, so both wild and domesticated animals can act as reservoirs of infection. In one of the very few studies on animal material, IS*6110* PCR and spoligotyping were used to examine a sample from a Pleistocene bison with an erosive lesion from the Wyoming Natural Trap Cave (Rothschild et al. 2001). However, the spoligotypes failed to match any in the database. Statistical analysis of the data indicated that the patterns had the closest match to *M. africanum*, although more recent analysis suggests ancestral *M. tuberculosis* (Huard et al. 2006). There was no indication of *M. bovis*. This is consistent with our current understanding of the evolution of the MTB complex, as we now know that *M. tuberculosis* is the more ancestral lineage (Brosch et al. 2002).

PCR based on IS*1081* is a better means of identifying *M. bovis*, as there are six copies per cell. Although earlier attempts to detect *M. bovis* in British Iron Age material (2,200 years BP) by this means were unsuccessful (Taylor et al. 2005), it has been found recently in a group of Siberian Iron Age semi-nomadic pastoralists (Taylor et al. 2007). However, its scarcity reflects the low levels of *M. bovis* infection reported in the absence of effective control measures today, of around 6% of cases (Grange 1995), and suggests that a long-term close association with an infected herd is the most likely scenario in which to detect this organism.

6.6.3 *Genotypes and Genetic Lineages*

It is clear that 'modern strains' of *M. tuberculosis*, defined as lineages with the TbD1 deletion, occurred in ancient Egypt, alongside more ancestral strains (Zink et al. 2003, 2005b). So far, mainly European and Near-Eastern archaeological samples have been subjected to further molecular analysis, and it would be of great interest to determine the nature of the indigenous tuberculosis found in archaeological material from pre-Colombian America, Africa, the Indian subcontinent and the Far East.

Further whole genome sequencing of modern clinical isolates has resulted in large international databases of the molecular characteristics of MTB strains, based on numbers of repetitive sequences, spoligotypes and SNPs. Recent meta-analyses of these databases has led to distinct lineages of *M. tuberculosis* strains being recognised, which are associated with different geographical regions and human populations (Baker et al. 2004; Hirsh et al. 2004; Gagneax et al. 2006), possibly contemporaneous with early hominids in Africa (Gutierrez et al. 2005).

6.6.4 *Tuberculosis and Ancestral Sequence Inference*

Our knowledge of the stability of molecular typing methods and their rate of change makes it possible to estimate the rate of evolutionary change under different scenarios. Meta-analyses of modern molecular data support the hypothesis that the other MTB complex species are clonally derived from an "*M. canettii*"-like organism, so this may offer a good genomic reference point to investigate how genes have evolved to greater virulence in *M. tuberculosis*. Analysis of one *M. tuberculosis* genotype (Beijing) has led to the hypothesis that it may have originated in central Asia in humans migrating from the Middle East during a second out-of-Africa migration in the Upper Palaeolithic 45,000–30,000 years ago (Mokrousov et al. 2005). In the future it may be possible to obtain direct evidence of MTB aDNA dating from these remote times, in order to explore the co-evolution of host and pathogen.

6.7 Interactions of *Mycobacterium tuberculosis*

6.7.1 Co-infections

It is important to appreciate that the relationship between human hosts and their microbial pathogens is dynamic and although in the case of tuberculosis the disease may remain latent for most of a lifetime, perturbations in the host cell-mediated immune response can lead to a re-activation of disease. Intestinal parasites, such as worms, have a profound effect upon host immunity, which can result in atopy rather than a cell-mediated protective response (Elston 2006). Other co-infections can also bring this about, and there is limited evidence in the archaeological record of individuals infected with both MTB and *Mycobacterium leprae* (Donoghue et al 2005), or *Leishmania* (Spigelman et al. 2005: Zink et al. 2006). The liver from a Korean 'wet' mummy (Kim et al. 2006) has been diagnosed with DNA from both MTB and hepatitis B virus (Donoghue et al. 2007).

6.7.2 Lowered Host Resistance or Increased Susceptibility

Host-related factors that can exacerbate the impact of MTB infection are extremes of age, nutritional stress, and neoplasms. Both the Vác and Nubian studies have shown a pronounced effect of age in relation to disease. The population at Kulubnarti in early Christian Nubia had a remarkably high incidence of the stress indicator *criba orbitaria*, which may be related to the high rates of infection and early deaths in the older settlement. One cause of *criba orbitaria* is iron deficiency anaemia, and in severe cases this enhances the virulence of MTB infection (Ratledge 2004). It has been suggested that the iron status within pre-Columbian populations in the Americas may have had a profound impact on the clinical presentation of the disease (Wilbur and Buikstra 2006). Work is just starting on the impact of neoplastic disease, but tuberculosis infection has been detected in an infant with Langerhans cell histiocytosis from the Vác mummy study group (Spigelman et al. 2006).

It is now becoming increasingly recognised that the genetic variability of *M. tuberculosis* strains has an impact on the clinical presentation of disease (Malik and Godfrey-Fausset 2005). A small proportion of strains currently cause a disproportionate number of cases of tuberculosis. Similarly, there is substantial evidence for the role of genetic factors in the susceptibility of humans to mycobacterial disease (Fernando and Britton 2006).

6.8 Conclusions

The palaeomicrobiology of tuberculosis has been illuminating to archaeologists, palaeopathologists, molecular epidemiologists and experts in microbial genomics alike. For the future, palaeomicrobiology offers us an exciting prospect of exploring the relationship between the microbial pathogen *Mycobacterium tuberculosis*, and its human host. This may enable us to examine directly different MTB strains and human genotypes, from a time before the selection pressure created by the global epidemic associated with the Industrial Revolution in the Western World.

Acknowledgements Palaeomicrobiology is impossible without the collaboration with palaeopathologists who supply specimens of interest. The independent verification of data relies upon colleagues who are willing to assist in this process. Therefore, my grateful thanks go to every one of my many collaborators around the world. In particular I acknowledge the long-term and continuing support of my colleague Mark Spigelman.

References

Abu Al-Soud W, Rådström P (2000) Effects of amplification facilitators on diagnostic PCR in the presence of blood, feces and meat. J Clin Microbiol 38:4463–4470

Armelagos GJ, Harper KN (2005) Genomics at the origins of agriculture, part two. Evol Anthropol 14:109–121

Armelagos GJ, Brown PJ, Turner B (2005) Evolutionary, historical and political economic perspectives on health and disease. Soc Sci Med 61:755–765

Arriaza BT, Salo W, Aufderheide AC, Holcomb TA (1995) Pre-Colombian tuberculosis in Northern Chile: molecular and skeletal evidence. Am J Phys Anthropol 98:37–45

Aufderheide A, Rodriguez Martin C (1998) The Cambridge encyclopedia of human paleopathology. Cambridge University Press, Cambridge

Baker L, Brown T, Maiden MC, Drobniewski F (2004) Silent nucleotide polymorphisms and a phylogeny for *Mycobacterium tuberculosis*. Emerg Infect Dis 10:1568–1577

Barnes P, Cave MD (2003) Molecular epidemiology of tuberculosis. N Engl J Med 349:1149–1156

Baron H, Hummel S, Herrmann B (1996) *Mycobacterium tuberculosis* complex DNA in ancient human bones. J Archaeol Sci 23:667–671

Boom R, Sol CJA, Salimans MMM, Jansen CL, Wertheim-van Dillen PME, van der Noordaa J (1990) Rapid and simple method for purification of nucleic acids. J Clin Microbiol 28:495–503

Braun M, Cook DC, Pfeiffer S (1998) DNA from *Mycobacterium tuberculosis* complex identified in North American, pre-Colombian human skeletal remains. J Archaeol Sci 25:271–277

Brosch R, Gordon SV, Marmiesse M, Brodin P, Buchrieser C, Eiglmeier K, Garnier T, Gutierrez C, Hewinson G, Kremer K, Parsons LM, Pym AS, Samper S, van Soolingen D, Cole ST (2002) A new evolutionary scenario for the *Mycobacterium tuberculosis* complex. Proc Natl Acad Sci USA 99:3684–3689

Brudey K, Driscoll JR, Rigouts L, Prodinger WM, Gori A, Al-Hajoj SA, Allix C, Aristimuño L, Arora J, Baumanis V, Binder L, Cafrune P, Cataldi A, Cheong S, Diel R, Ellermeier C, Evans JT, Fauville-Dufaux M, Ferdinand S, Garcia de Viedma D, Garzelli C, Gazzola L, Gomes HM, Guttierez MC, Hawkey PM, van Helden PD, Kadival GV, Kreiswirth BN, Kremer K, Kubin M, Kulkarni SP, Liens B, Lillebaek T, Ho ML, Martin C, Martin C, Mokrousov I, Narvskaïa O, Ngeow YF, Naumann L, Niemann S, Parwati I, Rahim Z, Rasolofo-Razanamparany V, Rasolonavalona T, Rossetti ML, Rüsch-Gerdes S, Sajduda A, Samper S, Shemyakin IG, Singh UB, Somoskovi A, Skuce RA, van Soolingen D, Streicher EM, Suffys PN, Tortoli E, Tracevska T, Vincent V, Victor TC, Warren RM, Yap SF, Zaman K, Portaels F, Rastogi N, Sola C (2006) *Mycobacterium tuberculosis* complex genetic diversity: mining the fourth international spoligotyping database (SpolDB4) for classification, population genetics and epidemiology. BMC Microbiology 6:23 http://www.biomedcentral.com/1471-2180-6-23

Butler WR, Guthertz LS (2001) Mycolic acid analysis by high-performance liquid chromatography for identification of *Mycobacterium* species. Clin Microbiol Rev 14:704–726

Canci A, Minozzi S, Borgognini Tarli SM (1996) New evidence of tuberculous spondylitis from Neolithic Liguria (Italy). Int J Osteoarchaeol 6:497–501

Cano RJ, Tiefenbrunner F, Ubaldi M, Del Cueto C, Luciani S, Cox T, Orkand P, Künzel KH, Rollo F (2000) Sequence analysis of bacterial DNA in the colon and stomach of the Tyrolean Iceman. Am J Phys Anthropol 112:297–309

Clark GA, Kelley MA, Hill MC (1987) The evolution of mycobacterial disease in human populations Curr Anthropol 28:45–62

Collins DM, Stephens DM (1991) Identification of an insertion sequence, IS*1081*, in *Mycobacterium bovis*. FEMS Microbiol Lett 67:11–15

Crubézy E, Ludes B, Poveda JD, Clayton J, Crouau-Roy B, Montagnon D (1998) Identification of Mycobacterium DNA in an Egyptian Pott's disease of 5,400 years old. C R Acad Sci III-Vie 321:941–951

Daniel TM (2000) The origins and precolonial epidemiology of tuberculosis in the Americas: can we figure them out? Int J Tuberc Lung Dis 4:395–400

Donoghue HD, Spigelman M (2006) Pathogenic microbial ancient DNA: a problem or an opportunity? Proc R Soc B 273:641–642

Donoghue HD, Spigelman J, Zias J, Gernaey-Child AM, Minnikin DE (1998) *Mycobacterium tuberculosis* complex DNA in calcified pleura from remains 1400 years old. Lett Appl Microbiol 27:265–269

Donoghue HD, Spigelman M, Greenblatt CL, Lev-Maor G, Kahila Bar-Gal G, Matheson C, Vernon K, Nerlich AG, Zink AR (2004) Tuberculosis: from prehistory to Robert Koch, as revealed by ancient DNA. Lancet Infect Dis 4:584–592

Donoghue HD, Marcsik A, Matheson C, Vernon K, Nuorala E, Molto JE, Greenblatt CL, Spigelman M (2005) Co–infection of *Mycobacterium tuberculosis* and *Mycobacterium leprae* in human archaeological samples: a possible explanation for the historical decline of leprosy. Proc R Soc B 272:389–394

Donoghue HD, Spigelman M, Grant P, Klein A, Shouval D, et al (2007) Microbiological investigation of a Korean medieval child mummy found in Yangju. In: Proceedings of the 6th World Congress on Mummy Studies, Teguise, Lanzerote, Canary Islands, Spain, 20–24 February 2007

Drancourt M, Raoult D (2005) Palaeomicrobiology: current issues and perspectives. Nat Rev Microbiol 3:23–35

Drobniewski FA, Caws M, Gibson A, Young D (2003) Modern laboratory diagnosis of tuberculosis. Lancet Infect Dis 3:141–147

Dziadek J, Sajduda A, Borun M (2001) Specificity of insertion sequence-based PCR assays for *Mycobacterium tuberculosis* complex. Int J Tuberc Lung Dis 5:569–574

Eisenach KD, Cave MD, Bates JH, Crawford JT (1990) Polymerase chain reaction amplification of a repetitive DNA sequence specific for *Mycobacterium tuberculosis*. J Infect Dis 161:977–981

Elston DM (2006) The hygiene hypothesis and atopy: bring back the parasites? J Am Acad Dermatol 54:172–179

Faerman M, Jankauskas R, Gorski A, Bercovier H, Greenblatt CL (1997) Prevalence of human tuberculosis in a Medieval population of Lithuania studied by ancient DNA analysis. Ancient Biomol 1:205–214

Fernando SL, Britton WJ (2006) Genetic susceptibility to mycobacterial disease in humans. Immunol Cell Biol 84:125–137

Fletcher HA, Donoghue HD, Holton J, Pap I, Spigelman M (2003a) Widespread occurrence of *Mycobacterium tuberculosis* DNA from 18th–19th Century Hungarians. Am J Phys Anthropol 120:144–152

Fletcher HA, Donoghue HD, Taylor GM, van der Zanden AGM, Spigelman M (2003b) Molecular analysis of *Mycobacterium tuberculosis* from a family of 18th century Hungarians. Microbiology 149:143–151

Forbes BA, Hicks KE (1996) Substances interfering with direct detection of *Mycobacterium tuberculosis* in clinical specimens by PCR: effects of bovine serum albumin. J Clin Microbiol 34:2125–2128

Formicola V, Milanesi Q, Scarsini C (1987) Evidence of spinal tuberculosis at the beginning of the fourth millennium BC from Arene Candide cave (Liguria, Italy). Am J Phys Anthropol 72:1–6

Fusegawa H, Wang BH, Sakurai K, Nagasawa K, Okauchi M, Nagakura K (2003) Outbreak of tuberculosis in a 2000-year-old Chinese population. Kansenshogaku Zasshi 77:146–149

Gagneux S, DeRiemer K, Van T, Kato-Maeda M, de Jong BC, Narayanan S, Nicol M, Niemann S, Kremer K, Gutierrez MC, Hilty M, Hopewell PC, Small PM (2006) Variable host-pathogen compatibility in *Mycobacterium tuberculosis*. Proc Natl Acad Sci USA 103:2869–2873

Garay SM (1996) Pulmonary tuberculosis In: Rom WN, Garay SM (eds) Tuberculosis. Little Brown, Boston, pp 373–412

Gernaey AM, Minnikin DE, Copley MS, Power JJ, Ahmed AMS, Dixon RA, Roberts CA, Robertson DJ, Nolan J, Chamberlain A (1998) Detecting ancient tuberculosis. Internet Archaeol 5: http://intarch.ac.uk/journal/issue5/gernaey_index.html

Gernaey AM, Minnikin DE, Copley MS, Dixon RA, Middleton JC, Roberts CA (2001) Mycolic acids and ancient DNA confirm an osteological diagnosis of tuberculosis. Tuberculosis (Edinb) 81:259–265

Gömez I Prat J, Mendonça de Souza SMF (2003) Prehistoric tuberculosis in America: adding comments to a literature review. Mem Inst Oswaldo Cruz 98(Suppl 1):151–159

Gordon SV, Heym B, Parkhill J, Barrell B, Cole ST (1999) New insertion sequences and a novel repeated sequence in the genome *of Mycobacterium tuberculosis* H37Rv. Microbiology 145:881–892

Götherström A, Collins MJ, Angerbjörn A, Lidén K (2002) Bone preservation and DNA amplification. Archaeometry 44:395–404

Grange JM (1995) Human aspects of *Mycobacterium bovis* infection. In: Thoen CO, Steele JH (eds) *Mycobacterium bovis* infection on animals and humans. Iowa State University Press, Ames, Iowa, pp 29–46

Gutierrez MC, Brisse S, Brosch R, Fabre M, Omaïs B, Marmiesse M, Supply P, Vincent V (2005) Ancient origin and gene mosaicism of the progenitor of *Mycobacterium tuberculosis*. PLoS Pathog 1:55–61

Haas CJ, Zink A, Molnár E, Szeimies U, Reischl U, Marcsik A, Ardagna Y, Dutour O, Pálfi G, Nerlich AG (2000) Molecular evidence for different stages of tuberculosis in ancient bone samples from Hungary. Am J Phys Anthropol 113:293–304

Haas F, Haas SS (1996) The origins of *Mycobacterium tuberculosis* and the notion of its contagiousness In: Rom WN, Garay SM (eds) Tuberculosis. Little Brown, Boston, pp 3–19

Hänni C, Brousseau T, Laudet V, Stehelin D (1995) Isopropanol precipitation removes PCR inhibitors from ancient bone extracts. Nucleic Acids Res 23:881–882

Hardie RM, Watson JM (1992) *Mycobacterium bovis* in England and Wales: past, present and future. Epidemiol Infect 109:23–33

Hirsh AE, Tsolaki AG, DeRiemer K, Feldman MW, Small PM (2004) Stable association between strains of *Mycobacterium tuberculosis* and their human host populations. Proc Natl Acad Sci USA 101:4871–4876

Hofreiter M, Serre D, Poinar HN, Kuch M, Pääbo S (2001) Ancient DNA. Nat Rev Genet 2:353–359

Höss M, Pääbo S (1993) DNA extraction from Pleistocene bones by a silica-based purification method. Nucleic Acids Res 21:3913–3914

Huard RC, Fabre M, de Haas P, Lazzarini LC, van Soolingen D, Cousins D, Ho JL (2006) Novel genetic polymorphisms that further delineate the phylogeny of the *Mycobacterium tuberculosis* complex. J Bacteriol 188:4271–4287

Huggett JF, McHugh TD, Zumla A (2003) Tuberculosis: amplification-based clinical diagnostic techniques. Int J Biochem Cell Biol 35:1407–1412

Hutás I (1999) The history of tuberculosis in Hungary. In: Pálfi G, Dutour O, Deák J, Hutás I (eds) Tuberculosis past and present. Golden Books/Tuberculosis Foundation, Budapest, pp 39–42

Kamerbeek J, Schouls L, Kolk A., van Agterveld M, van Soolingen D, Kuijper S, Bunschoten A, Molhuizen H, Shaw R, Goyal M, van Embden J (1997) Simultaneous detection and strain differentiation of *Mycobacterium tuberculosis* for diagnosis and epidemiology. J Clin Microbiol 35:907–914

Kemp BM, Monroe C, Smith DG (2006) Repeat silica extraction: a simple technique for the removal of PCR inhibitors from DNA extracts. J Archaeol Sci 33:1680–1689

Kim SB, Shin JE, Park SS, Bok GD, Chang YP, Kim J, Chung YH, Yi YS, Shin MH, Chang BS, Shin DH, Kim MJ (2006) Endoscopic investigation of the internal organs of a 15th-century child mummy from Yangju, Korea. J Anatomy 209:681–688

Konomi N, Lebwohl E, Mowbray K, Tattersall I, Zhang D (2002) Detection of mycobacterial DNA in Andean mummies. J Clin Microbiol 40:4738–4740

Lambert PA (2002) Cellular impermeability and uptake of biocides and antibiotics in Gram-positive bacteria and mycobacteria. J Appl Microbiol 92(Suppl):46S–54S

Mackowiak PA, Tiesler Bios V, Aguilar M, Buikstra JE (2005) On the origin of American tuberculosis. Clin Infect Dis 41:515–518

Malik ANJ, Godfrey-Fausset P (2005) Effects of genetic variability of *Mycobacterium tuberculosis* strains on the presentation of disease. Lancet Infect Dis 5:174–183

Manchester K (1984) Tuberculosis and leprosy in antiquity: an interpretation. Med History 28:162–173

Mark L, Patonai Z, Molnar E, Marcsik A (2006) High-throughput mass spectroscopic determination of ancient pathological biomarkers. In: Proceedings of the 16th Paleopathology Association European Meeting, 28 August–1 September 2006, Santorini Island, Greece

Mathema B, Kurepina NE, Bifani PJ, Kreiswirth BN (2006) Molecular epidemiology of tuberculosis: current insights. Clin Microbiol Rev 19:658–685

Mays S, Taylor GM, Legge AJ, Young DB, Turner-Walker G (2001) Paleopathological and biomolecular study of tuberculosis in a medieval skeletal collection from England. Am J Phys Anthropol 114:298–311

Mays S, Fysh E, Taylor GM (2002) Investigation of the link between visceral surface rib lesions and tuberculosis in a Medieval skeletal series from England using ancient DNA. Am J Phys Anthropol 119:27–36

Minnikin DE, Kremer L, Dover LG, Besra GS (2002) The methyl-branched fortifications of *Mycobacterium tuberculosis*. Chem Biol 9:545–553

Mokrousov I, Ly HM, Otten T, Lan NN, Vyshnevskyi B, Hoffner S, Narvskaya O (2005) Origin and primary dispersal of the *Mycobacterium tuberculosis* Beijing genotype: clues from human phylogeography. Genome Res 15:1357–1364

Nerlich AG, Haas CJ, Zink A, Szeimes U, Hagedorn HG (1997) Molecular evidence for tuberculosis in an ancient Egyptian mummy. Lancet 350:1404

Nuorala E, Götherström A, Ahlström T, Donoghue HD, Spigelman M, Lidén, K (2004) MTB complex DNA in a Scandinavian Neolithic passage grave. Paper I, Molecular Palaeopathology.

Ancient DNA analyses of the bacterial diseases tuberculosis and leprosy. In: Theses and Papers in Scientific Archaeology 6. Archaeological Research Laboratory, Stockholm University, Stockholm

O'Reilly LM, Daborn CJ (1995) The epidemiology of *Mycobacterium bovis* infections in animals and man: a review. Tuber Lung Dis 76 Suppl 1:1–46

O'Rourke DH, Hayes MG, Carlyle SW (2000) Ancient DNA studies in physical anthropology. Annu Rev Anthropol 29:217–242

Ortner DJ, Putschar WG (1981) Identification of pathological conditions on human skeletal remains. Smithsonian Institution Press, Washington DC

Pääbo S, Poinar H, Serre D, Jaenicke-Despres V, Hebler J, Rohland N, Kuch M, Krause J, Vigilant L, Hofreiter M (2004) Genetic analyses from ancient DNA. Annu Rev Genet 38:645–679

Pap I, Józsa L, Repa I, Bajzik G, Lakhani SR, Donoghue HD, Spigelman M (1999) 18–19th century tuberculosis in naturally mummified individuals (Vác, Hungary). In: Pálfi G, Dutour O, Deák J, Hutás I (eds) Tuberculosis past and present. Golden Books/Tuberculosis Foundation, Budapest, pp 421–428

Parsons LM, Brosch R, Cole ST, Somoskövi Á, Loder A, Bretzel G, Van Soolingen D, Hale YM, Salfinger M (2002) Rapid and simple approach for identification of *Mycobacterium tuberculosis* complex isolates by PCR-based genomic deletion analysis. J Clin Microbiol 40:2339–2345

Poinar HN, Stankiewicz BA (1999) Protein preservation and DNA retrieval from ancient tissues. Proc Natl Acad Sci USA 96:8426–8431

Poinar HN, Hofreiter M, Spaulding WG, Martin PS, Stankiewicz BA, Bland H, Evershed RP, Possnert G, Pääbo S (1998) Molecular coproscopy: dung and diet of the extinct Ground Sloth *Nothrotheriops shastensis*. Science 281:402–406

Raff J, Cook DC, Kaestle F (2006) Tuberculosis in the New World: a study of ribs from the Schild Mississippian population, West-Central Illinois. Mem Inst Owaldo Cruz 101[Suppl II]:25–27

Ratledge C (2004) Iron, mycobacteria and tuberculosis. Tuberculosis (Edinb) 84:110–130

Resnick D, Niwayama G (1995) Osteomyelitis, septic arthritis and soft tissue infection: organisms. In: Resnick D (ed) Diagnosis of bone and joint disorders. Saunders, Edinburgh, pp 2448–2558

Roberts CA, Buikstra JE (2003a) The development of tuberculosis from earliest times to the development of drugs. In: Davies PDO (ed) Clinical tuberculosis, 3rd edn. Arnold, London, pp 3–20

Roberts CA, Buikstra JE (2003b) The bioarchaeology of tuberculosis: a global view on a reemerging disease. University Press of Florida, Gainesville, FL

Rothschild BM, Martin LD (2003) Frequency of pathology in a large natural sample from Natural Trap Cave with special remarks on erosive disease in the Pleistocene. Reumatismo 55:58–65

Rothschild BM, Martin LD (2006) Did ice-age bovids spread tuberculosis? Naturwissenschaften 93:565–569

Rothschild BM, Martin LD, Lev G, Bercovier H, Bar-Gal GK, Greenblatt C, Donoghue H, Spigelman M, Brittain D (2001) *Mycobacterium tuberculosis* complex DNA from an extinct bison dated 17,000 years before the present. Clin Infect Dis 33:305–311

Salo WL, Aufderheide AC, Buikstra J, Holcomb TA (1994) Identification of *Mycobacterium tuberculosis* DNA in a pre-Columbian Peruvian mummy. Proc Natl Acad Sci USA 91:2091–2094

Santos AL, Roberts CA (2006) Anatomy of a serial killer: differential diagnosis of tuberculosis based on rib lesions of adult individuals from the Coimbra Identified Skeletal Collection, Portugal. Am J Phys Anthropol 130:38–49

Spigelman M, Donoghue HD (2003) Paleobacteriology with special reference to pathogenic mycobacteria. In: Greenblatt CL, Spigelman M (eds) Emerging pathogens: archaeology, ecology and evolution of infectious disease. Oxford University Press, Oxford, pp 175–188

Spigelman M, Greenblatt CL (1998) A guide to digging for pathogens. In: Greenblatt CL (ed) Digging for pathogens: ancient emerging diseases – their evolutionary, anthropological, and archaeological context. Balaban, Rehovot, Israel, pp 345–361

Spigelman M, Lemma E (1993) The use of the polymerase chain reaction (PCR) to detect *Mycobacterium tuberculosis* in ancient skeletons. Int J Osteoarchaeol 3:137–413

Spigelman M, Matheson C, Lev G, Greenblatt C, Donoghue HD (2002) Confirmation of the presence of *Mycobacterium tuberculosis* complex-specific DNA in three archaeological specimens. Int J Osteoarchaeol 12:393–400

Spigelman M, Greenblatt CL, Vernon K, Zylber MI, Sheridan SG, Van Gerven DP, Shaheem Z, Donoghue HD (2005) Preliminary findings on the paleomicrobiological study of 400 naturally mummified human remains from upper Nubia. J Biol Res 80:91–95

Spigelman M, Pap I, Donoghue HD (2006) A death from Langerhans cell histiocytosis and tuberculosis in 18th century Hungary – what palaeopathology can tell us today. Leukemia 20:740–742

Sreevatsan S, Escalante P, Pan X, Gillies DA 2nd, Siddiqui S, Khalaf CN, Kreiswirth BN, Bifani P, Adams LG, Ficht T, Perumaalla VS, Cave MD, van Embden JD, Musser JM (1996) Identification of a polymorphic nucleotide in *oxyR* specific for *Mycobacterium bovis*. J Clin Microbiol 34:2007–2010

Sreevatsan S, Pan X, Stockbauer KE, Connell ND, Kreiswirth BN, Whittam TS, Musser JM (1997) Restricted structural gene polymorphism in the *Mycobacterium tuberculosis* complex indicates evolutionarily recent global dissemination. Proc Natl Acad Sci USA 94:9869–9874

Steinbock RT (1976) Paleopathological diagnosis and interpretation: bone disease in ancient human populations. Thomas, Springfield, IL

Sterling TR, Pope DS, Bishai WR, Harrington S, Gershon RR, Chaisson RE (2000) Transmission of *Mycobacterium tuberculosis* from a cadaver to an embalmer. New Engl J Med 342:246–248

Sutlovic D, Gojanovic MD, Andelinovic S, Gugic, Primorac D (2005) *Taq* polymerase reverses inhibition of quantitative real time polymerase chain reaction by humic acid. Croat Med J 46:556–562

Tayles N, Buckley HR (2004) Leprosy and tuberculosis in Iron Age southeast Asia? Am J Phys Anthropol 125:239–256

Taylor GM, Crossey M, Saldanha J, Waldron T (1996) DNA from *Mycobacterium tuberculosis* identified in Mediaeval human skeletal remains using polymerase chain reaction. J Archaeol Sci 23:789–798

Taylor GM, Goyal M, Legge AJ, Shaw RJ, Young D (1999) Genotypic analysis of *Mycobacterium tuberculosis* from mediaeval human remains. Microbiology 145:899–904

Taylor GM, Mays S, Legge AJ, Ho TBL, Young DB (2001) Genetic analysis of tuberculosis in human remains. Ancient Biomol 3:267–280

Taylor GM, Stewart GR, Cooke M, Chaplin S, Ladva S, Kirkup J, Palmer S, Young DB (2003) Koch's Bacillus – a look at the first isolate of *Mycobacterium tuberculosis* from a modern perspective. Microbiology 149:3213–3220

Taylor GM, Young DB, Mays SA (2005) Genotypic analysis of the earliest known prehistoric case of tuberculosis in Britain. J Clin Microbiol 43:2236–2240

Taylor GM, Murphy E, Hopkins R, Rutland P, Chitov Y (2007) First report of *Mycobacterium bovis* DNA in human remains from the Iron Age. Microbiology 153:1243–1249

Van Gerven DP, Sandford MK, Hummert JR (1981) Mortality and culture change in Nubia's *Batn el Hajar*. J Hum Evol 10:395–408

Van Soolingen D (2001) Molecular epidemiology of tuberculosis and other mycobacterial infections: main methodologies and achievements. J Intern Med 249:1–26

Weed LA, Baggenstoss AH (1951) The isolation of pathogens from tissues of embalmed human bodies. Am J Clin Pathol 21:1114–1120

Weiss RA, McMichael AJ (2004) Social and environmental risk factors in the emergence of infectious diseases. Nature Med 10[Suppl]: 570–576

Wilbur AK, Buikstra JE (2006) Patterns of tuberculosis in the Americas – How can modern biomedicine inform the ancient past? Mem Inst Oswaldo Cruz 101[Suppl II]:59–66

World Health Organisation (2006) http://www.who.int/mediacentre/factsheets/fs104/en/index.html

Zink A, Haas CJ, Reischl U, Szeimies U, Nerlich AG (2001) Molecular analysis of skeletal tuberculosis in an ancient Egyptian population. J Med Microbiol 50:355–366

Zink AR, Reischl U, Wolf H, Nerlich AG (2002) Molecular analysis of ancient microbial infections. FEMS Microbiol Lett 213:141–147

Zink AR, Grabner W, Reischl U, Nerlich AG (2003a) Molecular study on human tuberculosis in three geographically distinct and time delineated populations from ancient Egypt. Epidemiol Infect 130:239–249

Zink AR, Sola C, Reischl U, Grabner W, Rastogi N, Wolf H, Nerlich AG (2003b) Characterization of *Mycobacterium tuberculosis* complex DNAs from Egyptian mummies by spoligotyping. J Clin Microbiol 41:359–367

Zink AR, Grabner W, Nerlich AG (2005a) Molecular identification of human tuberculosis in recent and historic bone tissue samples: the role of molecular techniques for the study of historic tuberculosis. Am J Phys Anthropol 126:32–47

Zink AR, Köhler S, Motamedi N, Reischl U, Wolf H, Nerlich AG (2005b) Preservation and identification of ancient *M. tuberculosis* complex DNA in Egyptian mummies. J Biol Res 80: 84–87

Zink A, Spigelman M, Schraut B, Greenblatt C, Nerlich A, Donoghue H (2006) Leishmaniasis in ancient Egypt and Upper Nubia. Emerg Infect Dis 12:1616–1617

Chapter 7
Past Leprae

Andreas G. Nerlich(✉) and Albert R. Zink

Abstract Although leprosy often results in characteristic morphological alterations to the skeleton, its diagnosis may be difficult in cases of less significant bone changes. Molecular analysis of such cases may help resolve several aspects of the palaeopathology and palaeoepidemiology of leprosy. Several reports have documented the extraction and molecular analysis of *Mycobacterium leprae* DNA from ancient bone samples. Accordingly, a direct palaeomicrobiological approach may be taken to investigate the disease and its sequelae. In addition, the origin and the spread of the disease, as well as the dramatic decline of this infection in post-mediaeval Europe, can now be investigated.

7.1 Introduction

Infectious diseases like tuberculosis and leprosy often result in characteristic morphological alterations to the skeleton, and thus can be identified easily in ancient human remains. However, in cases with less significant bone changes it can be more difficult to come to a clear diagnosis of the underlying disease. Especially in such cases, the analysis of genetic material in ancient tissues may help clarify an unsure morphological analysis. The recent development of modern molecular biological techniques, such as the polymerase chain reaction (PCR) and sequencing techniques, offers a new approach to the identification of pathogenic organisms (Zink et al. 2002; Drancourt and Raoult 2005). Such techniques not only help identify ancient bacterial DNA in human remains, thereby providing direct evidence of the occurrence and frequency of infectious diseases in historic populations, they also yield information about the evolution of microorganisms and the diseases they cause.

In this context, the molecular analysis of cases involving possible infection with *Mycobacterium leprae* is of particular interest, as several aspects of the palaeopathology

Andreas G. Nerlich
Institute of Pathology, Academic Hospital Munich-Bogenhausen, Englschalkinger Str. 77, 81925 Munich, Germany
E-mail: Andreas.Nerlich@extern.lrz.muenchen,de

D. Raoult and M. Drancourt (eds.), *Paleomicrobiology: Past Human Infections.*

and palaeoepidemiology of leprosy remain controversial (Aufderheide and Rodriguez-Martin 1998). Likewise, both the origin and the obvious spread of this disease, but also its dramatic decline in post-mediaeval Europe are unclear and require elucidation. To date, several reports have documented the extraction of *M. leprae* DNA from ancient bone samples (see below). Accordingly, a direct palaeomicrobiological approach may be taken to investigate this disease and its sequelae.

7.2 Clinical Aspects of Leprosy

Leprosy, or Hansen's disease, is a slowly progressive chronic infectious disease, caused by the bacillus *Mycobacterium leprae*, leading to granulomatous destruction of soft and hard tissues and potentially leading to severe mutilation of the infected individual. The disease was historically a major predator of mankind and – despite its present curability with specific antibiotics – ca. 500,000 individuals worldwide are still infected. Approximately 2–3 million people currently live with mutilations due to leprosy. Nowadays, most cases are concentrated in the tropics of South America, Africa and Asia, although sporadic endemic cases still occur in Europe (e.g. the Baltics, Eastern European countries), North America and the Pacific islands.

Steps in the transmission of the disease are not fully clear. However, it is accepted that the reservoir of the mycobacterium is exclusively human, and that it is most frequently transmitted by aerosolic spread of the bacilli. In most instances, infection seems to occur during childhood, with incubation times ranging from 6 months to several years. Rarely, unusually long incubation periods of up to 40 years have been reported (Gierke et al. 2000), although the infection rate in adults with close contact to infected individuals (e.g. spouses) is as low as 5% even on long-term investigation.

The clinical picture of leprosy is variable and depends on the type of host immune response. The course of the disease can be roughly divided into four stages, which may develop from one another.

Leprosy typically begins as an indeterminate form that can spontaneously heal, remain unchanged for a long time period or proceed to a more severe form. Approximately 95% of contacts with the bacillus will result in spontaneous resolution without development of clinical symptoms. This initial indeterminate form may produce ill-defined skin patches or maculae with slight hypopigmentation. In parallel, such patches may coincide with hypaesthesia of the corresponding skin nerve. If the disease progresses, tuberculoid leprosy may develop provided that the host immune response is still adequately preserved. In this stage, a rapid loss of skin sensation due to severe nerve damage may occur, as well as local paralysis, loss of sweat and sebaceous glands, and hair loss. The skin shows macular lesions with significant hypopigmentation; peripheral nerves are infiltrated and may present as thick subcutaneous bundles. Secondary symptoms include bruising of hypaesthesised skin due to local external damage, and superinfection with poorly healing ulcers.

The lepromatous stage occurs in individuals with a poor immune reaction. Clinically, this is the most severe form and can lead to disfiguring mutilation. The skin lesions may present as maculae, papulae or plaques with hypopigmentation. The regions most affected are ears, the central face, fingers and toes, but the distal extremities, e.g. the extensor surfaces of thighs and forearms, can also be affected. The severe infiltration of skin in the perinasal and periorbital region leads to the "facies leonine" or lion face, which is associated with loss of the eyelashes and lateral eyebrows ("facies leprosa"). Often, the eyes are affected causing blindness. Osseous resorption of the nasal aperture and destruction of the bridge of the nose result in severe mutilation of the face. Affection of the throat may lead to a typical hoarseness. In fact, all other body regions may also be affected leading to a variable clinical picture (see Sects. 7.4 and 7.6 for descriptions of osseous lesions).

A fourth stage, called the borderline stage, also exists, which is somewhat intermediate between the tuberculoid and the lepromatous stage in clinical symptoms.

Considering the wide range of clinical symptoms, especially the early and "milder" stages of the disease, leprosy can easily be confused with various other diseases. This is often important to reconcile in historical terms, since evidence previously interpreted as favouring a diagnosis of "leprosy" must be considered carefully. In contrast, the typical mutilations of the skeleton in the severe forms of leprosy leave such typical traces of the disease that it may be identified in historic remains with a high degree of certainty.

7.3 *Mycobacterium Leprae* – Molecular Features and Potential Typing

The infectious agent of leprosy, *Mycobacterium leprae*, belongs to the acid-fast bacilli group of mycobacteria, but has a number of particular features worthy of note. Like other species of the mycobacteriae group, *M. leprae* has a lipid-rich cell wall, which leads to the unusual staining properties of all acid-fact bacilli and which provides considerable protection to the bacillus. Thus, the conservation of *M. leprae* is much more likely than that of other bacteria in long-stored material, such as bone or mummified soft tissue from past populations.

On the other hand, on the genetic level the *M. leprae* bacillus is a somewhat "degenerated" mycobacterium since its genome has undergone significant downsizing and has accumulated more than 1,130 pseudogenes (Monot et al. 2005). As a consequence, the bacterium requires very particular growth conditions, and has a doubling time of as long as almost 13 days (Shepard and McRae 1965). *M. leprae* cannot be cultivated in in vitro cultures and the only systems available for the in vivo cultivation of the bacterium are the mouse pad model and the nine-banded armadillo *Dasypus novemcinctus* (Kirchheimer and Storrs 1971).

Extensive genetic analysis of the *M. leprae* genome – the entire length of which has recently been sequenced (Cole et al. 2001) – revealed extremely few differences between isolates from different regions of the world. Furthermore, there were no differences between strains from different sources (collected all over the world) on the level of the complete genome, the copy number of insertion-sequence-like dispersed repetitive sequences, including the mycobacterial interspersed repetitive unit (MIRU), and the variable number of tandem repeats (VNTR). Similarly, genetic fingerprinting and end-sequencing of numerous cosmids from a library of isolates with different origins showed perfect co-circularity between different strains. It was only on the level of single nucleotide polymorphisms (SNPs) that differences were noted (Monot et al. 2005). This latter study described an estimated overall frequency of SNPs in *M. leprae* of approximately one per 28 kb, which is significantly less than that observed in other human pathogens. These data strongly suggest that the *M. leprae* genome is exceptionally well conserved and that the leprosy bacillus is highly clonal (Smith et al. 1993).

The study of the worldwide distribution of SNPs in 175 specimens from 21 countries and all five continents identified only four different patterns, each with a distinct geographical distribution: type 1 occurs predominantly in Asia, the Pacific region and East Africa; type 4 is found in West Africa and the Caribbean region; type 3 resides in Europe, North Africa and the Americas; and, finally, type 2 (the rarest) is seen in Central/East Africa, North India/Nepal and New Caledonia. From this distribution, a general evolutionary scheme for *M. leprae* with two plausible scenarios has been derived. In the first scenario, SNP type 2 preceded type 1, spreading eastward from East Africa or Central Asia to East Asia and the Pacific region, and type 3 was disseminated westward to the Mediterranean and Central Europe before giving rise to type 4, which spread to America by colonialism. Alternatively, type 1 was the progenitor of type 2, followed by type 3 and finally type 4 (Monot et al. 2005).

Despite this recent breakthrough in strain identification patterning, the origin and the time axis of spread remains unclear. Likewise, it is uncertain if the origin of the disease lies in Central Africa or Central Asia; the route of spread is also an open debate. Indeed, these questions may be answerable only by palaeomicrobiological studies of relevant material from well-defined sources. Fortunately, leprosy in its full-blown clinical form leaves very typical traces in hard tissues, thus the analysis of human remains will probably provide adequate answers.

7.4 The Osteopathology of Leprosy

Since the palaeopathological record is restricted mostly to skeletal pathology, the specific and non-specific features of this disease will be outlined here in more detail. As indicated above, the advanced stage of leprosy is distinctive for the ailment, thus a diagnosis can be established with the necessary certainty. However, it is noteworthy that the indeterminate stages of the disease are not at all identifiable by bone pathology.

The typical osteopathology of leprosy was first described by Moller-Christensen (1961) in his superb analysis of the osseous remains from a leper cemetery in

Naestved, Denmark. Moller-Christensen described the typical alterations of the maxilla / nasal aperture and the small bones of the hands and feet (Moller-Christensen 1974). Concomitantly, the long bones of the distal limbs are also affected. However, the osseous pattern of these latter bones show non-specific alterations that are also seen in other chronic infectious diseases, such as tuberculosis or treponematosis, although to slightly differing degrees.

The skeletal involvement in leprosy ranges between 15 and 50% of affected individuals, although methodical examination of the skeletal populations of leprosaria indicates that almost 70% of such burials reveal leprosy-related skeletal alterations (Zimmermann and Kelly 1982; Moller-Christensen 1978). This concurs with modern leprosaria (Steinbock 1976); present day patients with leprosy have skeletal involvement in about 25% of cases (Paterson and Job 1964). Within this population, the most frequently affected body sites are the fingers and toes.

Since *M. leprae* affects nerves and other soft tissues along with direct skeletal affliction, skeletal lesions may be due to direct skeletal involvement, but may also result from secondary destruction due to infection of soft tissues. The latter may be particularly important in the destruction of fingers and toes, where loss of sensation (hypaesthesia) may result in secondary non-specific bacterial inflammation. Accordingly, skeletal involvement can be divided into two types:

1. Specific leprosy-induced skeletal changes include the so-called "rhinomaxillary syndrome" (Andersen and Manchester 1992) leading to the "facies leprosa" (Fig. 7.1). Furthermore, periostitis of long bones with subperiosteal new bone deposition occurs in more than 70% of leper cases (Moller-Christensen 1961). Most frequently, this is seen in tibiae and fibulae although other long bones may also be affected (lepromatous periostitis) (Fig. 7.2).
2. Non-specific inflammation and osseous degeneration occurs due to local trauma and secondary inflammation as a result of sensory loss. This osteitis/osteomyelitis is the same as that in patients without sensory loss and may lead to secondary periostitis, bone resorption and arthritis. These are the most frequent bone lesions found in the small bones of the hands and feet.

Finally, as a secondary effect – such as can result from chronic disuse – osteoporosis of skeletal segments may occur. The absence of periosteal reaction and callus formation in pathological fractures is very typically seen in leprosy (Schinz et al. 1953).

The horribly disfigured facial anatomy, known as "facies leprosa", is engraved in the skull bones as bilateral symmetrical resorption of the maxillary alveolar processes of the incisors with concomitant loss of the nasal aperture and formation of defects of the hard palate. Together, these processes lead to a wide and empty depression where the nose once existed. These bone changes are summarised as "rhinomaxillary syndrome" (Anderson and Manchester 1992). The syndrome – present only in lepromatous leprosy (and those borderline lesions close to lepromatous leprosy) – results from a direct involvement of the affected bones through *M. leprae* infection of the overlying mucosa and skin that spreads to the adjacent bone structures (Aufderheide and Rodriguez-Martin 1998). There is normally only little new bone formation at the periosteal surface, which represents an important differential

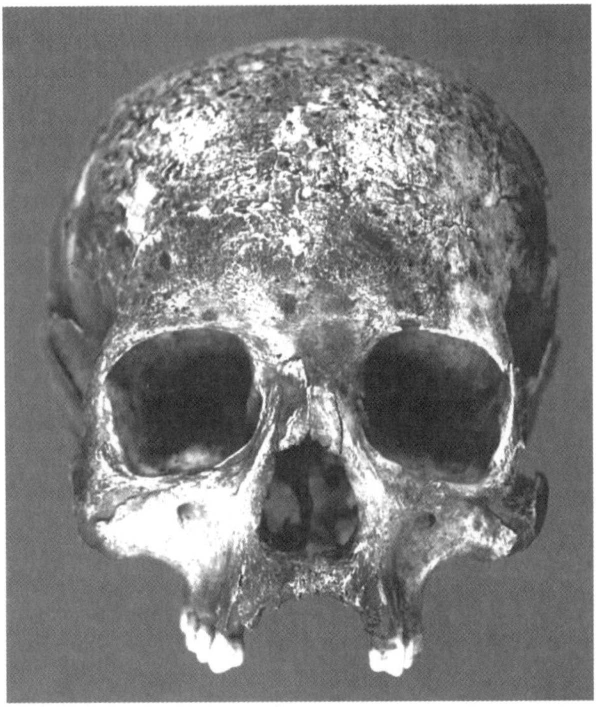

Fig. 7.1 Skull with typical pathological symptoms of 'facies leprosa' of the skull: wide aperture of the nose, extensive resorption of the maxilla and loss of the front teeth

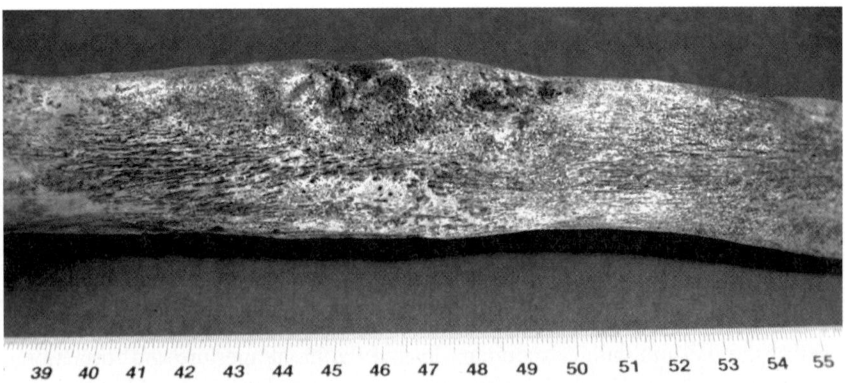

Fig. 7.2 Leprosy-associated osteitis / periostitis of the tibia showing enlargement of the complete bone with strongly irregular bone surface and focal resorption zones. *Scale* cm

diagnostic criterion in distinguishing facial inflammatory destruction not caused by leprosy (Revell 1986). The mandible is normally not affected.

The second body regions with direct skeletal involvement with the bacilli are the long bones, mainly those of the distal lower limbs. These show periostitis with subperiosteal new bone deposits. Such lepromatous periostitis, which is seen in up to 78% of leprosy cases (Moller-Christensen 1961), produces pitting and irregularity of the surface with fine, longitudinally striated, subperiosteal bone deposition. These features may be seen bilaterally, but may also occur unilaterally, most frequently in the distal third of the tibia, but also in fibulae, femorae and, rarely, in the long bones of the upper extremity.

In contrast, pathological changes in the hands and feet can be due to secondary destruction by non-specific inflammation due to small traumas along with the loss of sensation in the peripheral nerves. This can affect the proximal phalanges, the metacarpals and metatarsals with bilateral, though mostly asymmetric, infliction. The terminal phalanges concentrically erode down to a tapered appearance of the fingers. This deformation is also referred to as "licked candy stick" and finally leads to loss of individual phalanges or only stump-like fingers or toe-tips. In addition, the hands and feet of skeletons may show various degrees of dislocation.

About 15–50% of cases with verified leprosy show diagnostic skeletal pathology which, however, may be modified by superinfection (particularly of the small bones) with associated additional bone destruction. These traces may be uncovered during careful palaeopathological investigation and, accordingly, the certainty with which palaeopathological diagnosis can be made is influenced by the presence, and state of conservation, of the diagnostically relevant bones.

7.5 Literary and Iconographic Evidence of Leprosy in History

7.5.1 Origin and First Descriptions – From Early Dawn Until the Roman Period

The origin and spread of leprosy remain uncertain. The oldest evidence comes from literary and iconographic sources and therefore must be handled with great caution. In this regard, it is of particular significance that the clinical features of leprosy are very distinctive in its advanced stages, but are highly non-specific in its early indeterminate form. As a consequence, the interpretation of historic literary or artifactual evidence of leprosy may be difficult and ambiguous – especially if severe forms of the disease were rare or even absent.

Previously, numerous authors have associated the Hebrew word "tsara'ath" in the Old Testament (book of Leviticus) with leprosy (see Aufderheide and Rodriguez-Martin 1998). Since the Old Testament was probably written about 1500 B.C., "tsara'ath" has been regarded as the earliest written evidence of leprosy in antiquity. However, recent critical reviews raise serious concerns regarding the relationship between biblical descriptions of "tsara'ath" and leprosy. Although the diagnosis of "tsara'ath" was based on skin lesions with obvious hypopigmentation (and probably also with hypaesthesia), and resulted in an expulsion from human community, these leprosy-typical features must be reconsidered, since other symptoms and features mentioned in the Bible do not correlate well with leprosy. These concerns have been fuelled mainly by biblical records stating that “tsara’ath” was curable. Consequently, it is now more and more accepted that the biblical “tsara’ath” does not refer to leprosy sensu stricto, but rather to a broader range of various skin diseases (Marks 2002).

Similarly, it has become clear that the swelling skin disease described in the ancient Egyptian papyrus Ebers is much more likely to have been gas gangrene than leprosy. In summary, neither biblical nor ancient Egyptian texts provide sufficient evidence for the existence of leprosy at their respective times and regions.

As yet, the oldest reliable literary evidence for the disease can be dated back to the ancient literature of India. Medical texts dating to ca. 600 B.C. provide descriptions of certain features strongly suggestive of leprosy. Accordingly, in the early Indian textbook of “Sushruta Samhita” (Dharmendra 1947; Skinsnes 1973), the skin in leprosy (termed “Kuhthan”) is described as being “slightly vermillion-coloured, thin and spreading in its nature. A sort of pricking and piercing pain [is experienced in the affected locality] which loses all sensibility to the touches” (Marks 2002). This description is fairly consistent with a mild, tuberculoid form. Additionally, other clinical pictures of the disease are described, with the most extreme form exhibiting “contraction of the skin, local anaesthesia, a copious flow of perspiration, swelling, and piercing or cutting pain in the affected part together with a deformity of the limbs and hoarseness” (Marks 2002). Other symptoms include “breaking of the local skin…falling off the fingers,…sinking of the nose and ears and redness of the eyes” (Marks 2002). As Marks (2002) suggested, these symptoms are highly suggestive of the lepromatous form. Furthermore, the detailed description and classification of the disease into different stages suggests that it was fairly common in India at the time of the description. Further descriptions of similar pathological conditions are given in the book “Arthasastra”, which dates back to ca. 321–296 B.C. This book represents some sort of “manual on the art of government as a guide for kings and the maintainment of the earth” (Marks 2002). Here again, leprous conditions are mentioned along with suggestions for therapy by herbal medication. Marks (2002) suggests that the elaborate inclusion of leprosy into these guidelines strongly supports the presence of the disease at that time – and its presence some considerable time before.

In parallel, there is some evidence that leprosy might have been prevalent in China around 500 B.C. A Chinese document (attributed to Nei Ching Su Weng) attributes an illness to a historically known individual named Pai-Niu, a Confucian teacher. However, the specificity of this disease, termed “li”, as leprosy is still a matter of debate (Feeny 1964). Skinsnes (1980) cites a reference from a bamboo book dating back to ca. 250 B.C. that uses the term “li” for a disease characterised by nasal destruction, loss of eyebrows, crippling and fracturing of the legs, anaesthesia of the mucosa and hoarseness – very likely the lepromatous form of leprosy. Accordingly, this description has been regarded as strong support for the accuracy of the diagnosis in Pai-Niu’s case. In summary, there is literary evidence suggesting the presence of leprosy in India and China around 500 B.C., although as yet no skeletal – or even iconographic – evidence exists.

Artistic evidence of leprosy in the Greco–Roman period is also extremely uncertain, and an artefact regarded as a leprous hunchback by Grmek (1989) dating to 1300 B.C. and originating from Israel is unconvincing since the figurine fits much better to the dwarfed Egyptian god Bes, who was widely adored in the Egyptian empire.

In summary, literary and artifactual evidence for leprosy dates back to the ninth century B.C. in India and the sixth century B.C. in China. These observations suggest a descent of the disease from the Indian subcontinent (and/or ancient China). However, the origin of the disease is still uncertain.

7.5.2 *From the Roman Period Until the Late Middle Ages*

The next reliable literary descriptions of leprosy come from two Roman period historiographs: Celsus (25 B.C.–37 A.D.) and Aretaeus of Cappadocia (first century A.D.). Both provide relatively detailed descriptions of leprosy, which was termed "elephantiasis" but not "lepra" (Lechat 2002). Surprisingly, all Greek medical textbooks and all known historiographies before this time provide no description of the disease. This is of note, since, for example, the (significantly earlier) Corpus Hippocraticum contains numerous detailed descriptions of all kinds of contemporary diseases, but none is similar to, or even matches the symptoms of, leprosy (the Greek term "lepra" used by Hippocrates clearly relates to scaling of the skin, such as in psoriasis or fungal skin diseases). It is noteworthy that, almost in parallel, the Chinese surgeon Hua T'o provided a detailed description of a leonine face, thereby indicating that leprosy was still present in East Asia (ca. 150 A.D.; Aufderheide and Rodriguez-Martin 1998).

Pliny the Elder (23–70 A.D.) – a contemporaneous writer to Celus – also describes "elephantiasis" (meaning "leprosy" in our nomenclature) as having been brought to Rome by the returning army of Pompeius around 62 B.C., when he returned from a military campaign against the king of Pont, Mithridates (Lechat 2002). Although the exact attribution to leprosy remains uncertain, and its association with the army returning from Asia is also speculative, these descriptions of a disease possibly representing leprosy coming from the Middle East are of particular note. Recently, Lechat (2002) suggested that leprosy was at that time uncommon in the central Roman Empire since "elephantiasis" was not included in the list of diseases that was used to refuse the sale of slaves, as was the case for "phthisis" (tuberculosis), fevers, eyesores and mental disorders. However, in the period following, the spread of "elephantiasis" in the Roman Empire, e.g. to Gallia and Southern Germania, where Galenus describes a disease of leprous symptoms around 150 A.D., can be assumed (Lechat 2002).

An important observation with the spread of leprosy is the appearance at the beginning of the third century A.D. of special hospitals. These first leprosy hospitals (called "lazar houses") are recorded in Cappadocia and various countries in Central Europe (Ackerknecht 1972). This strongly supports the spread of the disease in the Roman Empire during that time. Furthermore, some descriptions suggest that socially high-ranking individuals, such as the Emperor Constantine, were also affected by leprosy. However, skeletal evidence and detailed literary description are still lacking.

In subsequent centuries, there is evidence for significant spread of the disease in Europe, but there is also continuous literary evidence for the disease in India and China. Likewise, the Chinese book "*Ch'ien Chin Yao Fang's One Thousand Golden Remedies*" describes typical features of leprosy and includes suggestions for some herbal medications (Skinsnes 1973). Meanwhile, the erection of lazar houses is increasingly documented in several European locations, such as England (638 A.D.) and Constantinople, but also in Japan (Wells 1964). Furthermore, leprosy was spread to Northern European countries by the Vikings, reaching Scandinavia in the tenth century. Here also, the disease affected individuals of various social classes. For example, in 1413, an Icelandic bishop was dismissed from his service since his leprosy-associated deformities prevented him from celebrating holy Mass (Lechat 2002).

The later medieval period is characterised by a continuous increase in the prevalence of leprosy in Europe, as evidenced by the number of lazar houses, until by the thirteenth century the presence of approximately 19,000 such special hospitals had been documented (Roberts 1986). The diagnosis of leprosy – as reported in various documents – was of great significance and was usually established by a special commission that contained specifically trained personnel, including infected members of lazar houses. As a consequence, it is not surprising that about 70% of leprosaria's occupants revealed the typical skeletal manifestations of leprosy (Moller-Christensen 1961). This strongly supports the concept that a diagnosis of leprosy was established carefully and was correct in a considerable percentage of suspected patients. Nevertheless, it is still a matter of strong debate whether the number of lazar houses indeed reflected infection rates by leprosy, and it has been claimed that the real prevalence of the disease was much lower than would be expected from the number of leprosy hospitals.

An important issue during this time period is the claimed association between the spread of leprosy and the crusades. It has repeatedly been hypothesised that leprosy was brought back by the knights of various crusades in the twelfth and thirteenth centuries. As yet, the only evidence for this hypothesis is the rapid concomitant increase in the number of lazar houses in Central Europe during this time period, suggesting increased disease prevalence. Without doubt, the Near Eastern region, including the Holy Land, was affected by leprosy, and there are even excellent descriptions of the disease affecting high-ranking persons, such as King Balduin of Jerusalem who died in 1185 at the age of 23 (Mitchell 2002). The "clinical" description of this case is very typical of his contracting borderline tuberculoid leprosy as a child, first noticed as an area of skin that had lost sensation on his right arm. With increasing age, he seems to have developed the lepromatous form of the disease, with typical mutilation, blindness and hoarse voice (Mitchell 2002). During that period, leprosy-infected crusaders founded a specially formed military order called the "Order of St. Lazarus", which enabled the infected to fight in the king's army despite being separated from the rest of the population.

Thus, although there is good evidence for leprosy in the crusader population, there is no proof for an active role of the crusades in the spread of the disease across Europe. Accordingly, although it is conceivable that many soldiers with signs and symptoms of leprosy took the disease with them on their return home having contracted it in the Near East, the already significant number of existing lepers in Central Europe makes it unlikely that they were the only ones to bring leprosy to Europe (Mitchell 2002).

7.5.3 From the Late Middle Ages Until Modern Times

Having reached a significant number in the Middle Ages, as mentioned above, a strong decrease in the number of lazar houses is noted by the sixteenth century. The reason for this remains an open question. Previously, Chaussinard (1948, 1953, 1966) suggested that a certain degree of cross-immunity between different mycobacteria caused reduced leprosy prevalence along with the increasing spread of tuberculosis. However, as yet there is no proof that the frequency of tuberculosis indeed increased considerably within the time frame in question. Furthermore, critical re-evaluation of the disease frequencies of leprosy and tuberculosis in modern day populations failed to reveal significant cross-interaction between these two diseases (Wilbur et al. 2002). Alternatively, it has been hypothesised that a novel strain of leprosy bacilli, which developed a much less aggressive clinical course, might have superseded the former strain. Finally, the separation of infected patients from the surrounding population might have led to a continuous decrease in the load of infectious sources, leading to a reduction in the number of new infections. However, why leprosy infections were selectively reduced, while other infectious diseases, such as tuberculosis, were not, remains unclear.

In Middle Europe, leprosy had disappeared almost completely by the end of the eighteenth century. However, endemic foci of the disease remained in Baltic and Scandinavian countries. Even today, isolated cases of leprosy occur in European countries, mostly imported from current hot spots where leprosy is endemic. However, in some cases the incubation periods seem to be extremely long and may have been missed upon superficial examination (Gierke et al. 2000).

7.5.4 Leprosy and the New World, Australia and Oceania

The Spanish conquest of Mesoamerica seems to have brought leprosy to the New World. At least there is no convincing evidence that the disease already existed in Pre-Columbian America. Similarly, the spread to the Pacific Islands seems to have been the result of European and/or Chinese colonisation. The first reference to leprosy in Hawaii was in 1823; not more than two generations later, almost 5% of the Hawaiian population suffered from leprosy (Ackerknecht 1972).

Recently, however, osteoarchaeological evidence has shaken this concept of modern day spread of leprosy in the Pacific area. Bone findings suggest that leprosy might have been present in Western Micronesia already between the seventh and fifteenth centuries A.D. (Trembly 1995, 2002), but may have been "overshadowed" by the later spread of leprosy during Western colonisation (see also Sect. 7.6).

7.6 Palaeopathological Findings in Leprosy Research

7.6.1 First Osteoarchaeological Evidence

Besides literary and iconographic evidence, the strongest evidence for leprosy comes from the methodical palaeopathological analysis of human remains, i.e. the bones surviving from burials at various places and from various time periods. As mentioned above, this holds true only for the lepromatous leprosy stages as only these produce the typical pathognomonic features of the disease that allow a concise diagnosis. All cases of the tuberculoid form will elude this type of analysis.

Currently, the oldest skeletal evidence of leprosy comes from a very recent palaeopathological analysis of a Celtic burial in Northern Italy, where Mariotti and co-workers (2005) identified a fourth–third century B.C. skeleton with some typical signs of leprosy, such as rhinomaxillary syndrome and typically malformed fingers. Archaeological evidence suggests that the adult male individual was a warrior who might have been involved in the Eastern Mediterranean wars and thus may have had contact to Near Eastern foci of leprosy. The authors speculate that leprosy spread rapidly to the Western world around the third–fourth century B.C. as single cases, apparently without producing an epidemic, since the affected skeleton was the only one out of 71 adults and 23 sub-adults.

The next skeletal evidence comes from the Ptolemeic (Greek) period in Egypt. In 1980, Dzierzykray-Rogalski described two skulls dating to approximately 200 B.C. found in the oasis of Dakhleh in the Western Desert, which demonstrated the typical lesions of facies leprosa. Recently, Molto (2002) described four further cases – also from the Egyptian desert oasis of Dakhleh and dated to the early-to-mid fourth century A.D. – with typical evidence of leprosy, here seen not only in the skulls but also in the typical malformations of the small bones of the fingers and toes. Covering only a short time period later, Wood-Jones (1908) described a further skull from a Nubian cemetery (fourth–seventh centuries) with destruction of the nasal bones, nasal septum and turbinates that also fits well with a diagnosis of leprosy; Moller-Christensen and Hughes (1966) reviewed and confirmed the diagnosis in this case. In addition, they identified a further skull from this Nubian series that also revealed signs of facies leprosa. A further early case dating to ca. 300–600 A.D. from Bet Guvrin in the Holy Land (Hershkovitz et al. 1992, 1993) was initially suggestive of leprosy, but on subsequent palaeomicrobiological analysis turned out to be a mixed infection with *M. leprae* ancient DNA along with non-specific inflammation (Spigelman and Donoghue 2001). A further case of leprosy from this region, however, was seen by Zias (1991, 2002) in seventh–tenth century material.

In parallel to this skeletal evidence for leprosy in the Near Eastern / Mediterranean region, first skeletal findings typical of leprosy in western European regions have been discovered in Britain, where a Romano-British skeleton from the fourth century A.D. presented with typical features of leprosy (Reader 1974). Further isolated cases suggestive of leprosy were described in a sixth century adult male skeleton from Gloucestershire (Wells 1962) and a seventh century male skeleton from Cambridgeshire (Moller-Christensen and Hughes 1962). A recent extensive survey of skeletal evidence of leprosy in Britain (Roberts 2002) on a total of 8,253 skeletons revealed 128 affected individuals. This survey covered 1,500 years, with 2 affected sites from the Romano-British period, 12 sites from the Anglo-Saxon period (fifth–eleventh centuries) and 27 sites from later periods (twelfth–seventeenth centuries). This suggests that there was an increase in leprosy over time, which correlates with the historical data. The first cases of leprosy in individual European countries have also been published as case reports: in France, two cases have been recorded from the Roman period of the fifth century (Blondiaux et al. 2002); the first case in Hungary was dated to 1082 A.D. (Palfi et al. 2002), in the Czech Republic 1293 A.D. (Dokladal 2002), and in Finland 1355 A.D. (Vuorinen 2002).

7.6.2 The Mediaeval Rise in Leprosy Prevalence

In parallel to the literary evidence outlined above, skeletal evidence of leprosy during the Mediaeval period is also increasing. Much information has come from extensive palaeopathological investigations of lazar house cemeteries – such as those performed by Moller-Christensen in the 1950s–1970s, and much more recently by Boldsen and co-workers (Boldsen 2001, 2005; Boldsen and Mollerup 2006). Such studies provide not only details of the typical osteopathological features of skeletons affected by leprosy, but form the basis for an estimation of the palaeoepidemiology of leprosy in distinct time periods. At present, this information is available only for Danish cemeteries, which have provided an excellent database for such estimates. However, one has to remember that leprosy infection rates in other countries – and also different time frames – may have been completely different.

In a first extensive palaeoepidemiological approach in 2001, Boldsen determined the rates of burials with signs of leprosy in three distinct settings. The Sanct Jorgen cemetery in Odense was the burial place of a lazar house and harboured 1,507 burials, of which 924 complete skeletons and 239 isolated skulls were present. This cemetery was in use between the thirteenth and the mid-seventeenth centuries. At least two-thirds of the people buried in this cemetery suffered from leprosy, which correlates well with previous findings by Moller-Christensen in a leper cemetery in Naestved, Denmark (1961). These data were compared with the findings in 200 adults from the cemetery of St. Jörgen in Malmö. Although this cemetery was in use between 1320 and 1520 A.D., the 200 burials under examination covered the late burial period (i.e. presumably after 1450 A.D.); not more than 10% of individuals were affected by leprosy. The third cemetery was that of a mediaeval village population from Tirup dating from the twelfth to the fourteenth century A.D. In the relatively small population of 61 adult skeletons analysed, ca. 35% of individuals showed features of leprosy. The frequency of leprosy in these three burial populations strongly suggests the following:

1. That leprosy was present in a significant proportion of the population in mediaeval Denmark, and was restricted not only to lazar houses, but also affected the rural population of small villages to a very considerable extent.
2. In later periods (ca. fifteenth/sixteenth centuries), the leprosy rates seem to have diminished considerably.
3. Both archaeological and literary evidence suggests that leprosy had disappeared from Denmark by the middle of the sixteenth century.

As a further interesting finding of this study, it turned out that the facial symptoms of leprosy (rhinomaxillary syndrome) were seen almost exclusively in the burials of lazar houses, while cases with minimal facial but more extensive peripheral osteopathology typical of leprosy dominated the leprosy cases in the Tirup sample. Furthermore, this study provides some evidence that people with leprosy symptoms died at a younger age than people without evidence of leprosy.

Subsequent studies by Boldsen (2005) and Boldsen and Mollerup (2006) determined the leprosy rates in four further cemeteries in central Denmark, covering various time frames between 1060 and 1818. All populations were of a considerable size, ranging between 66 and 372 well preserved adult skeletons. These populations revealed the prevalence of leprosy to have been between 13% and 23% in burials between 1060 and 1400, and 1–4% in material between 1200 and later than 1536. Accordingly, the prevalence of leprosy causing skeletal changes in the Early (1000–1200 A.D.) and High (1200–1400 A.D.) Middle Ages was very high, but was low in later burials. This independent study confirms the high prevalence of leprosy also in non-specialised cemeteries, thus confirming the aforementioned high prevalence of the disease found in skeletal remains from other sources.

7.6.3 The Post-Mediaeval Decline in Leprosy Frequency in Europe

A highly important phenomenon in the history of leprosy is the remarkable decline in the disease frequency in the post-mediaeval time period in Europe. This is evidenced both by the strong reduction in osteoarchaeological findings typical of leprosy and the considerable reduction in the number of leprosy hospitals. Thus, the number of "lazar houses" diminished after the fifteenth century. For example, in England, from a peak number of 200 lazar hospitals around the early fourteenth century, only very few were still recorded in the fifteenth and sixteenth centuries

(Manchester 1984). Similar figures have been reported from other countries and regions, suggesting a more general phenomenon. In only a few Northern European regions, such as western Norway and the Baltics, did leprosy remain an epidemic disease, maintaining a low-level prevalence rate in the population, until finally – following the identification in 1873, by the Norwegian doctor Amauer Hansen, of *M. leprae* as the infectious agent – the disease was extinct also in those regions (1955).

In parallel to the decline in the number of lazar hospitals, skeletal evidence also indicated that the rate of leprosy strongly declined in all European regions investigated. Detailed figures from the best analysed region to date – several cemeteries in Denmark (Boldsen 2001, 2005; Boldsen and Mollerup 2006) – were presented in Sect. 7.6.2.

In contrast to this decline in Europe, in other regions of the world a significant spread and increase in the disease has been noted, which parallels the literary evidence for the spread of leprosy in various regions (see also above).

7.6.4 Potential Reasons for the Extinction of Leprosy in Europe

The significant decline of leprosy in post-mediaeval Europe has been attributed to several factors, the effects of which, however, remain uncertain as yet. Currently, it is widely accepted that the segregation of lepers into lazar hospitals represents one important factor that reduced infection rates by the disease. However, taking into account the very long incubation periods (up to several years) and the high frequency of affected individuals during the peak incidence period (at least as documented by the "normal" village cemetery of Tirup, Denmark, with 25–50% leprosy-infected burials; Boldsen 2001), it is unlikely that the separation of the most severely affected lepers would have been sufficient to wipe out the disease.

As a further important factor, it has been claimed that leprosy-infected individuals were more susceptible to other epidemic diseases so that the great plague – which hit Middle Europe severely in 1348 and then repeatedly almost every 10–20 years – may have affected lepers more than the rest of the population. This may have led to a selective reduction in the number of lepers. Although this hypothesis is very interesting, it remains unclear why leprosy-infected individuals should have been affected more frequently, while at the same time the rate of tuberculosis infections increased. Tuberculosis obviously was not much affected by other epidemic diseases, although we have recently obtained molecular proof that the rate of tuberculosis infections was high in a group of plague victims (Zink et al. 2007).

As a further hypothesis, it has been claimed that cross-immunity between *Mycobacterium tuberculosis sive bovis* and *Mycobacterium leprae* might have led to the decline in leprosy, since increased tuberculosis rates seem to have paralleled the decline of leprosy (Chaussinard 1948, 1953, 1966). This interference hypothesis was based on the idea that while infection by *M. tuberculosis* offered some cross-immunity against leprosy, the converse was not true. As a consequence, tuberculosis wiped out leprosy. Experimental observations seem to support this idea. Additionally, this cross-immunity hypothesis has gained much support from epidemiological estimations (Lietman et al. 1997). Furthermore, the existence of cases with co-infection with leprosy and tuberculosis has been noted previously (Manchester 1984).

Recently, however, on the basis of recent endemic tuberculosis and leprosy data from Texas in the United States, Wilbur et al. (2002) suggested that the two diseases did not influence each other much, and that the rise and decline in one disease was paralleled by the same movement in the other. Very recently, in the largest molecular study on leprosy and tuberculosis to date, Donoghue et al. (2005) showed high levels of co-infection with both diseases in a selected population between the first and the fourteenth centuries, which was interpreted as a further indicator of an interaction between the two diseases. However, a recent large molecular study performed in our own laboratory on a mediaeval to modern day population from South Germany (Nerlich et al. 2007) found only a very low co-infection rate with both diseases. This issue will be discussed in more detail in Sect. 7.7.

In summary, the reason for the evident decline in leprosy around the fifteenth–sixteenth centuries remains as yet very unclear. Besides a multifactorial interaction involving changes in climate, separation of infected individuals, interfering epidemics with high mortality and potential cross-immunity, it may also be speculated that changes in *M. leprae* strains, with the appearance of strains with a much less aggressive clinical performance and concomitant "overgrowth" of the earlier *M. leprae*, might have led to the disappearance of the disease in most parts of Europe.

7.7 Analysis of Ancient *M. leprae* DNA

7.7.1 Methodological Remarks

The identification of *M. leprae* ancient DNA (aDNA) is facilitated by the fact that *M. leprae* has (like all bacilli of the mycobacteria group) an acid-fast cell wall that seems to protect the DNA from extensive diagenetic damage. Nevertheless, as for all aDNA studies, the target size is critical to any molecular analysis. For the specific identification of *M. leprae* DNA, different segments of the two repetitive elements RLEP1 and RLEP3 have most often been used for amplification by PCR (Yoon et al. 1993; Jamil et al. 1994), since these products are specific for *M. leprae* aDNA. Accordingly, PCR products of various size have been generated, in some cases surprisingly large fragments, e.g. in a study by Haas et al. (2000a), fragments of 372 bp and 320 bp were obtained for RLEP1 and RLEP3, respectively. Although fragments of this size may be criticised in terms of target length, both the aforementioned evident protection of the aDNA and the unambiguously positive results make RLEP1 and RLEP3 valuable targets in terms of aDNA research. Recently, Donoghue et al. (2001) identified and used primer pairs generating, on nested amplification, an outer amplification product of 136 bp and an inner product of 110 bp in length. Accordingly, this primer set covers a significantly smaller, but specific, *M. leprae* DNA segment and thus extends the possibilities available for aDNA research investigating leprosy. Indeed, own recent study on archival paraffin-embedded tissue material from a leprosy patient (which is a comparably "poorly" preserved historic tissue material) yielded a positive result with the Donoghue primer pair, but failed on both RLEP1 and RLEP3 amplifications (Nerlich et al. 2007).

7.7.2 Ancient DNA Analysis of Skeletal Remains – Reports from Isolated Cases or Small Series

The first successful molecular study on the identification of *M. leprae* was performed by Rafi et al. (1994a, 1994b) who positively identified *M. leprae* aDNA in the case of a seventh century leper from the Jordan River in Palestine. Using a protocol that investigated a 439 bp fragment of *M. leprae* DNA, they detected a specific positive amplification product in a severely destroyed first metatarsal bone. However, despite some clinical evidence of leprosy, two further samples in this series yielded negative results (Table 7.1).

The next report on the successful amplification of *M. leprae*-specific aDNA came from our own analysis of human remains from mediaeval- to modern-period skeletons (1400–1800 A.D.) from a small town ossuary in southern Germany (Haas et al. 2000a). Two skulls with typical rhinomaxillary syndrome, and therefore strongly suggestive of leprosy, tested unambiguously positive for both RLEP1 and

Table 7.1 Molecular results in ancient DNA (aDNA) leprosy research

Date (A.D.)	Number of positive cases	Provenance	Author	Publication year
First century	1	Israel	Donoghue et al.	2005
Fourth century	8	Dakhleh Oasis, Egypt	Donoghue et al.	2005
Fourth–seventh centuries	1	Israel	Spigelman and Donoghue	2001
Seventh century	1	Palestine	Rafi et al.	1994
Tenth century	1	Hungary	Haas et al.	2000
Tenth century	4	Hungary	Donoghue et al.	2001, 2005
Eleventh century	1	Hungary	Donoghue et al.	2005
Tenth–thirteenth centuries	1	Sweden	Donoghue et al.	2005
Mediaeval	1	Poland	Donoghue et al.	2001
Twelfth century	3	Spain	Montiel et al.	2003
Thirteenth–fourteenth centuries	1	Scotland	Taylor et al.	2000
Fifteenth–nineteenth centuries	5	South Germany	Haas et al.	2000
			Nerlich et al.	2007
Fifteenth century	1	Hungary	Donoghue et al.	2005

RLEP3 sequences. This was further confirmed by direct sequencing. In parallel, samples from two tenth-century burials from a Hungarian cemetery (Sarretudvari-Hizoföld) were investigated. While no leprosy-specific aDNA was amplifiable from the foot bones of either skeleton, the skull bone residues available from one individual tested positive. This study not only confirms the presence of leprosy in both populations, but suggests even more clearly that the bacillary load is significantly higher in rhinomaxillary lesions than in hand and foot bones with their presumed secondary infections.

Almost in parallel to the latter study, Taylor and co-workers (2000) reported a successful aDNA study on *M. leprae* in an individual from a Christian cemetery in the Orkney Islands, Scotland, dating to A.D. 1218–1370. The individual exhibited the typical rhinomaxillary features of severe lepromatous leprosy, and again aDNA was found only in skull bone samples, but not in those from other regions of the skeleton. The primer pair used covered a 153 bp segment of RLEP; the results were confirmed by direct sequencing.

Donoghue's highly specific nested primer pairs for the detection of leprosy (described above) have also been tested on archaeological material. Out of six samples, three, which were attributed to a nasal specimen from a mediaeval burial from Suraz, Poland, coming from a 40- to 50-year-old male with characteristic rhinomaxillary syndrome and severe mutilation of the fingers and toes, revealed positive amplification results. A positive amplification product was also found in a nasal specimen from two tenth–eleventh century Hungarian burials from Püspövladany, but no leprosy aDNA was seen in a metatarsal sample from another Hungarian cemetery (Opusztaszer-Monostor).

A further case of leprosy was identified by Spigelman and Donoghue (2001) in a 300–600 A.D. skeleton from Bet Guvrin, Israel, which presented with severe mutilation. Application of the Donoghue et al. (2001) primers in this case revealed a positive amplification result in a sample from the affected foot, thereby confirming leprosy in this individual.

The molecular analysis by Montiel et al. (2003) of skeletal remains in four adult skeletons from a twelfth century cemetery in Seville, Spain, showed a positive result for leprosy aDNA in three samples. In all cases, clinically affected metacarpal bone specimens were analysed using RLEP sequences generating 149 bp and 97 bp nested PCR products. This latter report was the first to show positive aDNA results in various members from an obvious leper community.

7.7.3 Ancient DNA Analysis of Skeletal Remains – Palaeoepidemiological Approaches

Following the above-listed reports on isolated cases or small series of molecularly proven leprosy, the first papers based on a molecular estimation of the palaeoepidemiology of leprosy in specific historic populations have started to appear. In part, these include previously published isolated cases, but some new cases are now included. Table 7.1 presents all the data available to date on molecularly identified cases of historic leprosy.

In 2005, Donoghue et al. presented positive molecular data on 16 cases of leprosy, 2 of which had previously been described by her group (Donoghue et al. 2001). Out of 30 additional individuals, 14 more positive leprosy aDNA cases were identified, covering a time period between the first century and the fourteenth–sixteenth centuries. The material came from Israel (first century, *n*=3, one leprosy positive); the Dakhleh Oasis, Egypt (fourth century; *n*=11; eight positive cases); Püspökladeny, Hungary (tenth century; *n*=5; four positive, of which two had been described previously); Szekesfehervar, Hungary (eleventh century; *n*=2, one positive); Björned, Sweden (tenth–thirteenth centuries; *n*= 3; one positive); Szekesfehervar, Hungary (fourteenth century; *n*=1, none positive); and Szombathely, Hungary (fifteenth century; *n*=3; one positive).

In this series, the rate of *M. tuberculosis* infection was molecularly tested in parallel in order to determine the rate of co-infection. Interestingly, a high frequency of *M. tuberculosis*-positive cases was also identified, with 18 positive cases out of a total number of 32 cases. Even more importantly, the rate of cases with co-infection was high, with 10 out of 24 cases revealing infection by both bacilli. Hence, the authors claim that co-infection (or even "superinfection" of leprosy by the more aggressive tuberculosis) caused an increased mortality rate in lepers, leading to the stepwise extinction of leprosy. This effect may have been aggravated by the socio-economic impact of segregation of leprosy patients, who were – at least in the serious clinical cases of lepromatous leprosy – readily identifiable by their facial mutilation (facies leprosa). Although this is the first study on a larger series of cases providing highly important and relevant data, it suffers from one major caveat: most leprosy cases in these study populations originated from lepers dating between the first and the tenth century, when the infection rate with leprosy was on an extreme incline (see Sects. 7.5 and 7.6), and not from the period when leprosy was wiped out at around the fifteenth–sixteenth century (only 4 of the 32 cases cover this time-period, with only one case testing positive for leprosy, two for tuberculosis, but none for co-infection). Accordingly, despite the significant value of this study, little can be concluded about the reduction in leprosy prevalence during the late Middle Ages and the beginning of modern times.

In order to potentially fill this gap, we have recently extended our own previous study on the molecular analysis of leprosy skulls (see Haas et al. 2000a) with a study on molecular leprosy identification in long bones with signs of chronic infection in a mediaeval to modern population dating from 1400 to 1800 A.D. (Nerlich et al. 2007). Out of a total population of at least 2,547 individuals (minimum individual number), 59 long bones with more-or-less clear morphological evidence of potential chronic infection were tested in parallel for the presence of *M. leprae* and *M. tuberculosis* aDNA. Sufficiently well preserved aDNA could be retrieved in 24 cases, with 10 cases containing *M. tuberculosis* DNA and 5 cases *M. leprae* aDNA (the latter included the two previously tested cases with rhinomaxillary lesions). Despite these significant infection rates for both mycobacterioses, only one case presented with co-infection.

This first methodical palaeopathological and molecular study analysed tuberculosis and leprosy by investigation of mycobacteria specific for tuberculosis and leprosy in the time period between the late Middle Ages and modern times (1400–1800 A.D.). Thereby, we provide evidence of significant infection by both infectious diseases in this population; however, the rate of co-infection in the study group was surprisingly low, thus this observation does not confirm the previously described high co-infection rate. Consequently, these first molecular observations do not support the idea that tuberculosis "wiped out" leprosy due to its more aggressive and destructive growth pattern. Moreover, it is conceivable that, after a period of (more-or-less peaceful) co-existence between leprosy and tuberculosis over ten centuries, either the leprosy strain or the environmental conditions for leprosy changed significantly leading to a reduction in the disease frequency. Finally, this recent study does not lend support to the previous cross-immunisation hypothesis proposed by Chaussinard and others, but takes into account rather more the critical observations of Wilbur et al. (2002). Nevertheless, until a novel proof for any hypothesis arises, this issue remains to be clarified.

7.8 Conclusions and Perspectives

Ancient DNA research and palaeomicrobiology have opened new debates about the origin, spread, and disappearance of leprosy in Europe, as well as in other regions of the world. In this regard, it is important to remember that although certain infectious diseases can manifest with characteristic pathological bone alterations, clinically milder infections – such as the early indeterminate or even the tuberculoid types of leprosy, will remain unidentified by such means.

At present, numerous molecular studies have identified *M. tuberculosis* in various tissue samples from diverse regions and different time periods (e.g. Salo et al. 1994; Nerlich et al. 1997; Haas et al. 2000b; Zink et al. 2003a, 2003b, 2004, 2005; Donoghue et al. 2004, 2005). Far less data exist on *M. leprae*, although several protocols have been established for the successful amplification and identification of its DNA in ancient bone samples. Along with this increasing knowledge, the first studies providing molecularly proven insights into the spread and prevalence of the disease are beginning to appear. However, considering the findings of Boldsen (Boldsen 2001, 2005; Boldsen and Mollerup 2006), which suggest a very high prevalence rate of leprosy in leper communities, but also in "normal" village burials (up to 25–50% of individuals affected), the two most recent molecular studies contain only a few and obviously very selected cases. Consequently, these preliminary data, though valuable, do not at all reflect the "clinical" reality of mediaeval leprosy.

Nevertheless, first insights into basic data on leprosy are emerging from different sources. Literary, osteoarchaeological and comparative molecular analysis of recent *M. leprae* strains from different countries worldwide strongly suggest that the disease originated in Central Africa, India and/or Central China, with subsequent spread westward (and possibly eastward). The advent of literary and palaeopathological evidence of the disease in the Mediterranean region around 300–500 B.C. suggests possible spread of the disease by warfare or commercial exchange. However, the apparently low prevalence rates at that time may suggest a "less harmful" bacillus or a more favourable host–pathogen interaction between humans and the mycobacteria at that time. This may also be reflected in the surprisingly high co-infection rates with leprosy and tuberculosis.

The pattern reveals significant changes during the Middle Ages, with almost an explosion of infections, together with specifically targeted measures to control the disease (in special hospitals or "lazar houses"), and the occurrence of concurrent epidemics of highly lethal bacilli such as the Black Death. The obvious increase in the numbers of infected persons may have been the consequence of either a novel and more aggressive bacillus strain (as yet unidentified) or a weakened host–pathogen reaction.

The reason for the dramatic decrease in the disease in the fifteenth–sixteenth centuries in Central Europe – despite its persistence in isolated Northern European spots – is also not clear at present and deserves further investigation. Both recent epidemiological (Wilbur et al. 2002) and molecular studies raise serious concerns regarding the hypothesis that cross-immunisation between *M. tuberculosis* strains and *M. leprae* may have been the reason for this decline. Other mechanisms, such as a novel change in the leprosy bacillus strain pattern or other features may be more plausible, although as yet unproven.

Accordingly, the molecular investigation of *M. leprae* in historic tissue material is now, more than 10 years after the first successful palaeomicrobiological identification (Rafi et al. 1994a, 1994b), still in its infancy. Ongoing studies are urgently required to shed more light on the palaeobiology of this unusual pathogen, which was (particularly in the pre-antibiotic era) one of the biggest predators of mankind.

References

Ackerknecht EH (1972) History and geography of the most important diseases. Hafner, New York

Andersen JG, Manchester K (1992) The rhinomaxillary syndrome in leprosy: a clinical radiological and paleopathological study. Int J Osteoarcheol 2:121–129

Aufderheide AC, Rodriguez-Martin C (1998) The Cambridge encyclopedia of human paleopathology. Cambridge University Press, Cambridge, UK, pp 141–154

Blondiaux J, Dürr J, Khouchaf L, Eisenberg LE (2002) Microcopic study and X-ray analysis of two 5th century cases of leprosy: paleoepidemiological inferences. In: Roberts CA, Lewis ME, Manchester K (eds) The past and present of leprosy. Archaeological, historical, paleopathological and clinical approaches. Archaeopress, Oxford, pp 105–110

Boldsen JL (2001) Epidemiological approach to the paleopathological diagnosis of leprosy. Am J Phys Anthropol 115:380–387

Boldsen JL (2005) Leprosy and mortality in the medieval Danish village of Tirup. Am J Phys Anthropol 126:159–168

Boldsen JL, Mollerup L (2006) Outside St. Jorgen: leprosy in the medieval Danish city of Odense. Am J Phys Anthropol 130:344–351

Chaussinard R (1948) Tuberculose et lèpre, maladies antagoniques. Int J Lepr 16:431–438

Chaussinard R (1953) Tuberculosis and leprosy: mutually antagonistic diseases. Lepr Rev 24:90–94

Chaussinard R (1966) Quelques remarques concernant la théorie de l'antagonism entre tuberculose et lèpre. Int J Lepr 28:484–485

Cole ST, Eiglmeier K, Parkhill J, James KD, Thomson NR, Wheeler PR, Honoré N, Garnier T, Churcher C, Harris D, Mungall K, Basham D, Brown D, Chillingworth T, Connor R, Davies RM, Devlin K, Duthoy S, Feltwell T, Fraser A, Hamlin N, Holroyd S, Hornsby T, Jagels K, Lacroix C, Maclean J, Moule S, Murphy L, Oliver K, Quail MA, Rajandream MA, Rutherford KM, Rutter S, Seeger K, Simon S, Simmonds M, Skelton J, Squares R, Squares S, Stevens K, Taylor K, Whitehead S, Woodward JR, Barrell BG (2001) Massive gene decay in the leprosy bacillus. Nature 409:1007–1011

Dharmendra R (1947) Leprosy in ancient Indian medicine. Int J Lepr 15:424–430

Dokladal M (2002) The history of leprosy in the territory of the Czech Republic. In: Roberts CA, Lewis ME, Manchester K (eds) The past and present of leprosy. Archaeological, historical, paleopathological and clinical approaches. Archaeopress, Oxford, pp 155–156

Donoghue HD, Holton J, Spigelman M (2001) PCR primers that can detect low levels of *Mycobacterium leprae* DNA. J Med Microbiol 50:177–182

Donoghue H, Spigelman M, Greenblatt CL, Bar-Gal GK, Lev-Maor G, Matheson C, Vernon K, Nerlich AG, Zink AR (2004) Ancient DNA from the *Mycobacterium tuberculosis* complex and *Mycobacterium leprae* in archaeological material – verification and characterization. Lancet Infect Dis 4:584–592

Donoghue HD, Marcsik A, Matheson C, Vernon K, Nuorala E, Molto JE, Greenblatt CL, Spigelman M (2005) Co-infection of *Mycobacterium tuberculosis* and *Mycobacterium leprae* in human archaeological samples: a possible explanation for the historical decline of leprosy. Proc Biol Sci 272:389–394

Drancourt M, Raoult D (2005) Palaeomicrobiology: current issues and perspectives. Nat Rev Microbiol 3:23–35

Dzierzykray-Rogalski T (1980) Paleopathology of the Ptolemaic inhabitants of Dakhleh oasis (Egypt). J Human Evol 9:71–74

Feeny P (1964) The fight against leprosy. Elek, London

Gierke U von, Stagelschmidt P, Prantl F, Strian F, Flaschenträger T (2000) Seltene Ursache eines Karpaltunnelsyndroms: Lepra. MMW Fortschr Med 142:39–40

Grmek M (1989) Diseases in the ancient Greek world. Johns Hopkins University Press, Baltimore, MD

Haas CJ, Zink A, Palfi G, Szeimies U, Nerlich AG (2000a) Detection of leprosy in ancient human skeletal remains by molecular identification of *Mycobacterium leprae*. Am J Clin Pathol 114:428–436

Haas CJ, Zink A, Molnar E, Szeimies U, Reischl U, Marcsik A, Ardagna Y, Dutour O, Palfy G, Nerlich AG (2000b) Molecular evidence for different stages of tuberculosis in ancient bone samples from Hungary. Am J Phys Anthropol 113:293–304

Hansen GA (1955) [Causes of leprosy] translation from the Norwegian 1873 original. Int J Lepr 23:307–309

Hershkovitz I, Spiers M, Matznelson A, Arensberg B (1992) Unusual pathological conditions in the lower extremities of a skeleton from ancient Israel. Am J Phys Anthropol 88:23–26

Hershkovitz I, Spiers M, Arensberg B (1993) Leprosy or Madura foot? Ambigous nature of infectious disease in paleopathology. Am J Phys Anthropol 91:251–253

Jamil S, Wilson SM, Hacket M, Hussain R, Stoker NG (1994) A colorimetric PCR method for the detection of *M. leprae* in skin biopsies from leprosy patients. Int J Lepr 62:512–520

Kirchheimer WF, Storrs EE (1971) Attempts to establish the armadillo (*Dasypus novemcinctus* Linn.) as a model for the study of leprosy. I. Report of lepromatoid leprosy in an experimentally infected armadillo. Int J Lepr 39:693

Lechat MF (2002) The paleoepidemiology of leprosy: an overview. In: Roberts CA, Lewis ME, Manchester K (eds) The past and present of leprosy. Archaeological, historical, paleopathological and clinical approaches. Archaeopress, Oxford, pp 157–162

Lietman T, Porco T, Blower S (1997) Leprosy and tuberculosis: the epidemiological consequences of cross-immunity. Am J Public Health 87:1923–1927

Manchester K (1984) Tuberculosis and leprosy in antiquity: an interpretation. Med Hist 28:162–173

Mariotti V, Dutour O, Belcastro MG, Facchini F, Brasili P (2005) Probable early presence of leprosy in Europe in a Celtic skeleton of the 3rd–4th century BC (Casalecchio di Reno, Bologna, Italy). Int J Osteoarcheol 15:311–325

Marks S (2002) Alexander the Great, seafaring and the spread of leprosy. J Hist Med 57:285–311

Mitchell PD (2002) The myth of the spread of leprosy with the crusades. In: Roberts CA, Lewis ME, Manchester K (eds) The past and present of leprosy. Archaeological, historical, paleopathological and clinical approaches. Archaeopress, Oxford, pp 171–178

Moller-Christensen V (1961) Bone changes in leprosy. Munksgaard, Copenhagen

Moller-Christensen V, Hughes DR (1962) Two early cases of leprosy in Great Britain. Man 287:177–179

Moller-Christensen V, Hughes DR (1966) An early case of leprosy from Nubia. Man 1:242–243

Moller-Christensen V (1974) Changes in the anterior nasal spine and alveolar process of maxilla in leprosy. Int J Lepr 42:431–435

Moller-Christensen V (1978) Leprosy changes of the skull. Odense University Press, Odense

Molto JE (2002) Leprosy in Roman period skeletons from Kellis 2, Dakhleh, Egypt. In: Roberts CA, Lewis ME, Manchester K (eds) The past and present of leprosy. Archaeological, historical, paleopathological and clinical approaches. Archaeopress, Oxford, pp 179–192

Montiel R, Garcia C, Canadas MP, Isidro A, Gujio JM, Malgosa A (2003) DNA sequences of *Mycobacterium leprae* recovered from ancient bones. FEMS Microbiol Lett 226:413–414

Monot M, Honore N, Garnier T, Araoz R, Coppée JY, Lacroix C, Sow S, Spencer JS, Truman RW, Williams DL, Gelber R, Virmond M, Flageul B, Cho SN, Ji B, Paniz-Mondolfi A, Convit J, Young S, Fine PE, Rasolofo V, Brennan PJ, Cole ST (2005) On the origin of leprosy. Science 308:1040–1042

Nerlich AG, Haas CJ, Zink A, Szeimies U, Hagedorn HG (1997) Molecular evidence for tuberculosis in an ancient Egyptian mummy. Lancet 350:1404

Nerlich AG, Marlow S, Zink A (2007)Ancient DNA analysis of leprosy and tuberculosis in a medieval to modern skeletal series from a distinct South German population. In: Proceedings of the 8th International Conference on Ancient DNA and Associated Biomolecules, Łódź, Poland, 16–19 October 2006. Medical University of Łódź (in press)

Palfi G, Zink A, Haas CJ, Marcsik A, Dutour O, Nerlich AG (2002) Historical and paleopathological evidence for leprosy in Hungary. In: Roberts CA, Lewis ME, Manchester K (eds) The past and present of leprosy. Archaeological, historical, paleopathological and clinical approaches. Archaeopress, Oxford, pp 205–212

Paterson D, Job C (1964) Bone changes and absorption in leprosy: a radiological, pathological and clinical study. In: Cochrane R, Davey T (eds) Leprosy in theory and practice, 2nd edn. Wright, Bristol, pp 425–446

Rafi A, Spigelman M, Stanford J, Lemma E, Donoghue H, Zias J (1994a). *Mycobacterium leprae* DNA from ancient bone detected by PCR. Lancet 343:1360–1361

Rafi A, Spigelman M, Stanford J, Lemma E, Donoghue H, Zias J (1994b) DNA of *Mycobacterium leprae* detected in ancient bone. Int J Osteoarchaeol 4:287–290

Reader R (1974) New evidence for the antiquity of leprosy in early Britain. J Archaeol Sci 1:205–207

Revell PA (1986) Pathology of bone. Springer, Berlin

Roberts C (1986) Leprosy and leprosaria in mediaeval Britain. MASCA J 4:15–21

Roberts CA (2002) The antiquity of leprosy in Britain – skeletal evidence. In: Roberts CA, Lewis ME, Manchester K (eds) The past and present of leprosy. Archaeological, historical, paleopathological and clinical approaches. Archaeopress, Oxford, pp 213–221

Salo WL, Aufderheide AC, Buikstra J, Holcomb TA (1994) Identification of *Mycobacterium tuberculosis* DNA in a pre-Columbian Peruvian mummy. Proc Natl Acad Sci USA 91:2091–2094

Schinz HR, Baensch WE, Friedl E, Uehlinger E (1953) Enfermedades inflammatorios de los huesos. In: Schinz HR, Baensch WE, Friedl E, Uehlinger E (eds) Röntgen-Diagnostico. Salvat, Barcelona

Shepard CC, McRae DH (1965) *Mycobacterium leprae* in mice: minimal infectious dose, relationship between staining quality and infectivity, and effect of cortisone. J Bacteriol 89:365–379

Skinsnes OK (1973) Notes from the history of leprosy. II. Chronology of selected books on leprosy. Int J Lepr 41:234–237

Skinsnes OK (1980) Leprosy in archaeologically recovered bamboo book in China. Int J Lepr 48:333

Smith JM, Smith NH, O'Rourke M, Spratt BG (1993) How clonal are bacteria? Proc Natl Acad Sci USA 90:4384

Spigelman M, Donoghue HD (2001) Unusual pathological condition in the lower extremities of a skeleton from ancient Israel. Am J Phys Anthropol 114:92–93

Steinbock RT (1976) Paleopathological diagnosis and interpretation: bone disease in ancient human populations. Thomas, Springfield, IL

Taylor GM, Widdison S, Brown IN, Young D (2000) A medieval case of lepromatous leprosy from 13–14th century Orkney, Scotland. J Archaeol Sci 27:1133–1138

Trembly DL (1995) On the antiquity of leprosy in Western Micronesia. Int J Osteoarchaeol 5:377–384

Trembly DL (2002) Perspectives on the history of leprosy in the Pacific. In: Roberts CA, Lewis ME, Manchester K (eds) The past and present of leprosy. Archaeological, historical, paleopathological and clinical approaches. Archaeopress, Oxford, pp 233–238

Vuorinen HS (2002) History of leprosy in Finland. In: Roberts CA, Lewis ME, Manchester K (eds) The past and present of leprosy. Archaeological, historical, paleopathological and clinical approaches. Archaeopress, Oxford, pp 239–246

Wells C (1962) A possible case of leprosy from a Saxon cemetery at Beckford. Med Hist 6:383–386

Wells C (1964) Bones, bodies and disease. Thames and Hudson, London

Wilbur AK, Buikstra JE, Stojanowski C (2002) Mycobacterial disease in North America: an epidemiological test of Chaussinard's cross-immunity hypothesis. In: Roberts CA, Lewis ME, Manchester K (eds) The past and present of leprosy. Archaeological, historical, paleopathological and clinical approaches. Archaeopress, Oxford, pp 247–258

Wood Jones F (1908) The pathological report. Archaeol Survey Nubia Bull 2:55–69

Yoon K-H, Cho S-N, Lee M-K, Abalos RM, Cellona RV, Fajardo TT Jr, Guido LS, Dela Cruz EC, Walsh GP, Kim J-D (1993) Evaluation of polymerase chain reaction amplification of *Mycobacterium leprae*-specific repetitive sequence in biopsy specimens from leprosy patients. J Clin Microbiol 31:895–899

Zias J (1991) Leprosy and tuberculosis in the Byzantine monasteries of the Judean desert. In: Ortner DJ, Aufderheide AC (eds) Human paleopathology. Current synthesis and future options. Smithsonian Institute Press, Washington DC

Zias J (2002) New evidence for the history of leprosy in the ancient Near East: an overview. In: Roberts CA, Lewis ME, Manchester K (eds) The past and present of leprosy. Archaeological, historical, paleopathological and clinical approaches. Archaeopress, Oxford, pp 259–268

Zimmerman MR, Kelly MA (1982) Atlas of human paleopathology. Praeger, New York

Zink A, Reischl U, Wolf H, Nerlich AG (2002) Molecular analysis of ancient microbial infections. FEMS Microbiol Lett 213:141–147

Zink A, Grabner W, Reischl U, Wolf H, Nerlich AG (2003a) Molecular study on human tuberculosis in three geographically distinct and time delineated populations from ancient Egypt. Epidemiol Infect 130:239–249

Zink AR, Sola C, Reischl U, Grabner W, Rastogi N, Wolf H, Nerlich AG (2003b) Characterization of *Mycobacterium tuberculosis* complex DNAs from Egyptian mummies by spoligotyping. J Clin Microbiol 41:359–367

Zink AR, Sola C, Reischl U, Grabner W, Rastogi N, Wolf H, Nerlich AG (2004) Molecular identification and characterisation of *Mycobacterium tuberculosis* complex in ancient Egyptian mummies. Int J Osteoarchaeol 14:404–413

Zink A, Grabner W, Nerlich AG (2005) Molecular identification of human tuberculosis in recent and historic bone tissue samples. A study on the role of molecular techniques for the study of historic tuberculosis. Am J Phys Anthropol 126:32–47

Zink AR, Signoli M, Ardagna Y, Maczel M, Dutour O, Nerlich AG (2007) Molecular evidence for tuberculosis in vertebral bone tissue of burials of a plague cemetery. In: Dutour O et al (eds) Proceedings of the ICEPID4. Plague: epidemics and societies (in press)

Chapter 8
Archaeology of Human Pathogens: Palaeopathological Appraisal of Palaeoepidemiology

Olivier Dutour

Abstract The recent introduction of the new field of research of palaeomicrobiology has established new links between microbiological and archaeological sciences by using molecular techniques on archaeological material. However, although the material under study appears to be shared by both these fields, some of the methods, concepts, expectations and paradigms are not. The goal of this chapter is to present, from the bioanthropological and palaeopathological point of view, what ancient bones can tell us concerning the reconstruction of past infectious diseases from a palaeoepidemiological perspective.

8.1 Introduction: the Evolutionary Paradigm

The general framework of the history of human pathogens is inscribed into the evolutionary paradigm – scientifically introduced in its modern form by Charles Darwin in 1859 (Darwin 1859)[1]. This paradigm is necessary and sufficient to explain, in the field of human infection, phenomena such as the extinction of human diseases ('suette', *lues maligna praecox*, Spanish flu), the appearance of new ones (AIDS, Legionnaire's and Mad Cow diseases), and the re-emergence of others [tuberculosis (TB)]. Based on this strong paradigm, it has been possible to build models of co-evolution, clearly illustrated by the Reed Queen Theory (Van Valen 1973), that have been invaluable to the understanding of host–pathogen interactions (Combes 2001).

From this perspective, the possibility of accessing primary data (i.e. human remains) allows us to examine the history of human infections more directly, in order to better understand their present-day evolution over a longer time scale, and to reexamine the phenomenon of re-emergence in terms of its real evolutionary significance.

Olivier Dutour
Visiting Professor: University of Toronto Department of Anthropology,
19 Russel Street, Toronto, ON M5S 2S2, Canada
E-mail: oj.dutour@utoronto.ca

[1]Surprisingly, despite its fundamental importance for all living species, this paradigm is still disputed (directly or insidiously) at the beginning of this new millennium.

D. Raoult and M. Drancourt (eds.), *Paleomicrobiology: Past Human Infections.*

Palaeopathology, palaeoradiology and palaeomicrobiology (Drancourt and Raoult 2005), including palaeoparasitology, thus play a key role in the retrospective diagnosis of infectious diseases in ancient human remains.[2]

8.2 From Epidemiology to Palaeoepidemiology

The tentative leap from retrospective diagnosis (by analogy in palaeopathology and palaeoradiology, or demonstratively in palaeomicrobiology) to global evaluation of pathological conditions in past populations is attempted by the field of palaeoepidemiology (for a general introduction, see Cohen and Crane-Kramer 2003). Palaeoepidemiology can be defined as the '*Use of epidemiologic methods to infer how certain diseases might have been distributed in ancient times; how, why and where they originated and, armed with this information, to predict possible futures of communicable and other diseases, possible trends in the emergence of new diseases, and reemergence of old ones. Evidence comes from contemporary accounts and from archaeological studies (evidence derived from bones, teeth, stomach contents)*' (MediLexicon 2007).

As this definition refers to methods of investigation in epidemiology, it is of interest to summarise these. After collecting data on a population (more often on a statistically representative sample of the population under study), modern descriptive epidemiology can calculate incidence and prevalence rates (Gerstman 2003). Prevalence measures the total number of cases of a disease in a given population; incidence corresponds to the rate of occurrence of new cases in this population. It should be noted that *Incidence* (i.e. number of new cases of a disease during a given time interval) is often used to mean *Incidence rate* (incidence divided by the number of people at risk, often expressed as the incidence per 1,000). Incidence can be called the 'absolute risk' (AR), and incidence rate the 'relative risk' (RR). Thus, incidence rate provides information about the risk of contracting the disease, whereas prevalence is a measure of how common the disease is (typically expressed as a percentage). It is also of interest to distinguish two types of prevalence: *point* and *period* prevalence. Point prevalence measures the proportion of people in a population who have a disease at a particular time; it represents a 'snapshot' of the disease in time. Period prevalence evaluates the proportion of people in a population who have the disease over a specific period of time (e.g. a season or a year). Period prevalence is distinct from incidence, because it concerns *all* affected individuals, (regardless of the date of contraction); whereas incidence concerns only those individuals who have *newly* contracted the disease during the same specified time interval.

Lifetime prevalence is the number of individuals (expressed as a percentage) in a statistical population that, compared to the total number of individuals, have

---

[2] Although generally not considered as palaeopathology or palaeomicrobiology research (probably because of its applied consequences) the 'resurrection of the Spanish flu virus' clearly falls into this category.

experienced the disease at some point in their life (up to the time of assessment). Lifetime morbidity risk is the theoretical prevalence of a disease at any point in life for anyone, regardless of time of assessment.

Duration of a disease influences both prevalence and incidence: a disease with a long duration may have a high prevalence but a low incidence rate; a disease with a short duration (but easily transmitted) may have a low prevalence but a high incidence rate. In other words, prevalence is a useful criterion for evaluation of long-lasting diseases, but incidence is a more relevant parameter when discussing diseases of short duration.

These parameters are well defined and are commonly used in clinical epidemiology; the diachronic approach opens fundamental new perspectives on our knowledge of the evolution of disease. Thus epidemiology can inform palaeoepidemiology. However, some clouds obscure the blue serenity of this sky. Measuring disease frequency in the past is far from the relatively straightforward procedure it is in present populations (Waldron 1994; Dutour et al. 1998, 2003). The main reason is that past populations are represented mainly by skeletal populations, and these do not properly represent past populations as they existed when alive. In fact, skeletal series are the worst type of sample for an epidemiologist. Paradoxically, it could be stated that palaeoepidemiology has very little to do with the epidemiology of past populations as it concerns the epidemiology of skeletal samples only.

It is obvious that when describing disease frequency in palaeopathology, some of the measures, such as incidence rate, commonly used by modern epidemiologists are impossible to attempt. Appropriate rates that can be used in palaeoepidemiology include prevalence, period prevalence, proportional morbidity rates, and age-specific prevalence rates (Waldron 1994). More recently, Boldsen (2001) defined 'point prevalence at death', and suggested that this rate could be obtained from a formula using sensitivity and specificity rates.

Considering prevalence, we can examine how the numerator and denominator differ in palaeoepidemiology from the modern situation, in order to establish, if possible, more relevant comparisons of past and present infectious conditions. What we must appreciate is what the ratio n/N (in which 'n' is the number of palaeopathological cases of a given disease observed in ancient human remains, and 'N' the number of individuals constituting the archeological population sample under study) really represents.

The 'n' question refers to palaeopathological diagnosis, and the 'N' question to the nature of the sample (represented by a collection of human remains).

8.3 Palaeopathological Diagnosis: the '*n*' Question

According to Brothwell (1961), in palaeopathology '…diagnosis is by far the greatest problem'. This is due to the specificities of the subject, including (1) the retrospective diagnosis, (2) the use of modern diagnosis criteria, (3) the scarcity of pathognomonic lesions, and (4) the incomplete nature of ancient material. In a concept introduced by some physicians such as William Osler, in modern medical practice the patient is a collective of signs and symptoms to be characterised and analysed algorithmically in order to reach a diagnosis. This process of identifying a pathological condition is based on a set of diagnostic criteria, including a spectrum of various types of information and observations as well as the results of different investigations. This includes anamnesis, complete examination and complementary analyses (medical imaging, lab tests, etc.). In modern medicine, diagnosis may be achieved using analogical (e.g. association of symptoms such as sub-acute or chronic asthaenia, vesper fever, weight loss, and radiological thoracic opacity could indicate several diseases, among them tuberculosis) and/or demonstrative (e.g. PCR analyses demonstrating the presence of *Mycobacterium tuberculosis*) procedures. In palaeopathology, retrospective diagnosis is silent (no anamnesis, no medical history), static (no evolution of signs and symptoms) and limited (mainly to skeletal expression). Indeed, mummified tissues are exceptional – the majority of ancient human remains are represented only by bones and teeth. This means that many diseases are under-represented or completely lacking because they leave no, or only minimal, imprints on bone, and many diseases that do affect bones may be confused with each other as they do so in a similar manner. In addition, many diseases can cause death before enough time has elapsed for bone to be affected (Ubelaker 1998).

Natural processes (physical, chemical, and biological) – so-called taphonomic processes (Mays 1992) – acting upon ancient skeletal remains will interfere with the '*n*' question of diagnosis in two ways: firstly the preservation state of the skeletal material, which can be fragmentary, incomplete or intermingled, will influence the quality of observations; secondly, taphonomic alterations can mimic disease conditions and induce interpretation errors (so-called pseudopathology), sometimes even for experienced palaeopathologists.

Two other comments should be added to complete the picture of the '*n*' question for palaeopathological diagnosis. The first is the 'attraction force' of the typical form of a disease; the second is 'forgotten diagnoses'. Palaeopathology has, as did medicine in early modern times, focussed its interest on 'typical' cases. Much as biological anthropology did in the first half of the twentieth century, by studying individual 'type' rather than population variability, palaeopathology at that time was interested mainly in 'casuistic reasoning' rather than in the actual 'health status' of past populations. From a certain point of view, to base a diagnosis on only the typical expression of a disease is reasonable. On other hand, as we know from clinical experience, diseases are rarely, if ever, represented by a single typical symptom, but rather by a set of major and minor signs; thus the scoring of only 'pathognomonic changes' will underevaluate the past prevalence of a given disease. If we consider the theoretical '*n*' as the sum of pathognomonic changes (n_0) and other minor symptoms representing various clinical expression of the same disease ($n_1+n_2+n_3$....), assessing only the n_0/N rate will clearly minimise the real presence of the disease in the past population under study. The n_0/N rate should be interpreted as the 'minimal' prevalence of the disease. The practical example of tuberculosis clearly illustrates this point. From the palaeopathological point of view, only the typical skeletal changes of Pott's disease are reliable for retrospective TB diagnosis, with precise diagnostic criteria: involvement of one to four vertebrae in the same area, destructive lesions, vertebral collapse producing angular kyphosis, posterior involvement uncommon, and anterior concavity of several adjacent vertebrae corresponding to the presence of a cold abscess (Aufderheide and Rodriguez-Martin 1998; Ortner 2003). However, other extraspinal skeletal involvements due to TB are also frequent (osteoarthritis of joints, especially hip and knee; osteomyelitis of long and short bones, especially femora, tibia, and foot bones) and may represent the only skeletal lesions attributable to a tuberculous infection in a palaeopathological specimen. Moreover, some minor palaeopathological changes, such as rib internal lesions (Santos and Roberts 2006) or endocranial serpiginous lesions (Hershkovitz et al. 2002; Schultz 1999) have recently been correlated with TB infection, as confirmed by using palaeomicrobiological techniques (Maczel et al. 2005). Taking into account all of the less typical changes associated with skeletal TB (extra-spinal and minor changes) will strongly modify estimates of TB prevalence. For example, tuberculosis changes were scored on 1,294 Hungarian skeletons from the medieval–modern period: the prevalence of tuberculosis estimated using only typical changes is about 0.2%; this rises to 3.8% (about 20 times more) if all skeletal expressions of TB are considered (Maczel, 2003).

The second comment concerns 'forgotten diagnoses'. In 1888, the French physician Victor Ménard published a book in which he summarised the courses given by Professor Lannelongue at the Faculty of Medicine in Paris (Ménard 1888). In his book, he pointed out that the term 'vertebral tuberculosis' refers not only to the 'classic' form known as Pott's disease[3], where the typical vertebral collapse can be seen, but should also include other manifestations: superficial 'carious' lesions or 'superficial vertebral tuberculous osteoperiostitis' (Fig. 8.1). He distinguished the two anatomical forms of vertebral tuberculosis (classical Pott's and superficial vertebral lesions) by the fact that they might appear separately. He pointed out that superficial vertebral 'caries' are frequently associated with visceral lesions; vertebral lesions are characterised by the lack of reparation, and the affected individuals frequently die of TB. The extension of these superficial lesions, appearing as small excavations on the anterior surface and lateral sides of vertebrae, is often considerable (they generally affect 5–6 to 12 vertebrae). The denuded surface shows variable aspects: sometimes it is smooth and plain, but generally it is rough, irregular, mined by small sinuous excavations, covered at the sides by newly formed bone layers, and infiltrated by 'fungosity'.

[3] It should be noted that Sir Percival Pott described, in 1779, 'morbific alterations' of vertebrae of unclear origin, and that, in 1816, Jacques-Mathieu Delpech proposed that these lesions be called 'tuberculous infection of vertebrae', pointing out the fact that this was the first time that this disease had been assigned a characteristic name: *L'état de la science sur ce point est tel qu'il convient aujourd'hui d'appeler cette maladie infection tuberculeuse des vertèbres, et ce sera la première fois qu'elle aura reçu une dénomination caractéristique* (J.M. Delpech 1816)

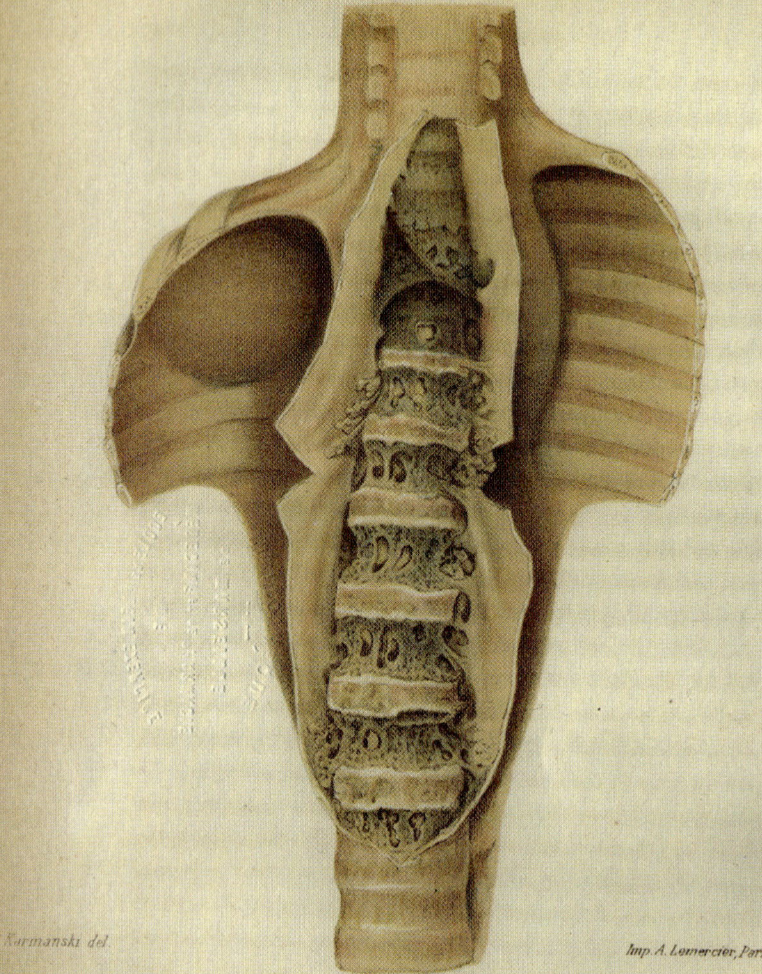

TUBERCULOSE SUPERFICIELLE ÉTENDUE A UN GRAND NOMBRE DE CORPS VERTÉBRAUX (CARIE DES ANCIENS AUTEURS); MAL DE POTT SURAJOUTÉ.

On voit à la surface des vertèbres des érosions, de petites ulcérations, des cavités superficielles, les unes vides, les autres pleines de fongosités et de pus. Quelques disques intervertébraux sont diminués de hauteur et n'ont plus une forme régulière. Un vaste abcès tuberculeux prévertébral est en rapport avec ces altérations multiples du squelette. Sous la plèvre pariétale droite proémine un abcès tuberculeux symptomatique d'une altération de la troisième côte. (V. Obs. XXXVIII, p. 402)

Fig. 8.1 Anatomical lesions described by Lannelongue as "superficial vertebral tuberculous osteoperiotitis" (Ménard 1888). Reproduced courtesy of the Library of the University of la Mediterranée, Collection of Ancient Medical Books, Faculty of Medicine of Marseille

The infection is suggested to progress along the blood vessels entering the vertebra, represented by the enlargement of vascular channels (see details in Maczel 2003). This description has totally disappeared from modern literature on skeletal tuberculosis – it was last mentioned by Sorrel and Sorrel-Dejerine (1932). This could mean either that this clinical expression of TB no longer exists in modern populations or that, because of its scarcity or difficulties in observing such features by medical imaging, it is ignored by modern clinicians. The re-discovery of these vertebral lesions by palaeopathologists is quite recent. Baker (1999) suggested that the 'smooth walled resorptive lesions/severe circumferential pitting', observed in some vertebral columns of four osteoarcheological series, might be of tubercular origin. Her hypothesis, ignoring Ménard's description, was based on the co-occurrence of these vertebral changes with other pathological conditions indicating TB. Among osteological collections with known cause of death, a frequent association has been found between these vertebral lesions and tuberculosis, especially in younger age groups (Pàlfi et al. 2000[4]; Ortner 2003). Haas et al. (2000) were the first to use molecular techniques to establish the relationship between these superficial vertebral alterations and tuberculosis.

Such 'forgotten diagnoses' should remind us that (1) old clinical descriptions are interesting, (2) the natural expression of infectious diseases is strongly influenced by our modern preventive and curative arsenal, and (3) modern clinical diagnostic criteria are, consequently, not the most appropriate way to establish diagnoses of infectious diseases in old bones.

In order that, as put by Waldron (1994), the attempt to establish retrospective diagnosis in palaeopathology will not become as difficult as 'trying to navigate through a minefield with the aid of the sun and a Mickey Mouse watch', we would do well to bear these points in mind.

8.4 The Nature of the Sample: the '*N*' Question:

A skeletal 'population' is in fact a sub-sample of several other samples. Of course, the sampling is not randomised. The main extrinsic factors contributing to the constitution of osteoarchaeological series are (1) burial assemblage (influenced by cultural practices), (2) duration (time of constitution of the sample, sometimes extending over several centuries), (3) taphonomic processes (chemical or biological), and (4) condition of the archaeological excavation.

[4] Palaeopathology of tuberculosis. Contribution to the knowledge of the evolution of the disease. Oral presentation by Pálfi Gy, Dutour O, Ortner DJ given in Budapest at the 1st European Region Conference of the International Union Against Tuberculosis and Lung Disease, 12–15 April 2000.

8.4.1 Burial Assemblage

Burial assemblage, which can influence the sample structure (Sellier 1996) and consequently the reconstruction of prevalence of diseases, depends on cultural practice. For example, some ancient civilisations buried their children separately (Watts 1989; Blaizot et al. 2003). If the disease under palaeoepidemiological study presents age-specific prevalence rates (which is, for example, the case for tuberculosis), it will be of interest to determine if the youngest individuals are missing or under-represented in the skeletal series because of specific burial practices.

In other cases, burial practices concern gender selection (e.g. monastic cemeteries), where the skeletal material obviously displays a very specific age and sex distribution (for instance mainly old men), inducing an over- or under-estimation of age-/sex-specific prevalence rates of some diseases (Waldron 1985). Thus, the burial assemblage must be precisely known in order to define the skeletal sample; the ideal sample is a non-selected population.

8.4.2 Time Effect

As Waldron suggested, period prevalence seems to be the most adapted rate in palaeoepidemiology – the period frequently being a very long one (Waldron 1994).

Large skeletal samples frequently come from excavations of a hypogea, necropolis, or cemetery that was in use over several centuries. In such cases, the skeletal population is the sum of the dead portions of the successive living populations. Although it is usually difficult to date the burials archaeologically, precisely separating the chronological sub-samples is easier (Boldsen 2001), and the 'population' is defined more by the burial place than by its chronological range. It is unlikely that a population would have remained static in structure and origin over a period of several centuries (except in the uncommon case of a genetically isolated community with a stable economic status); however, this bias of heterogeneity remains outwith our control (Wood et al. 1992).

The reconstructed prevalence of diseases, which can be single or recurrent events, chronic or acute, will generally be minimised, tending to a mean prevalence for the total period, especially for acute or sporadic phenomena (Dutour et al. 2003). Even if the prevalence cannot really be predicted, it must be taken into consideration when studying the epidemiology of skeletal series. The shorter the period involved in the constitution of the skeletal sample, the better the sample for palaeoepidemiology.

8.4.3 Taphonomy

The effect of taphonomy on palaeoepidemiology is twofold. On a general level, a poor state of preservation of a skeletal sample will reduce its interest for palaeoepidemiology. Preservation indices (Dutour 1989; Bello 2000) can quantify the overall preservation state of a skeletal series and provide information on the intensity of taphonomic processes. Clearly, the number of individuals alone is insufficient to prescribe the material available for palaeoepidemiological studies, and each skeletal population presents its own general preservation profile. On a more detailed level, differences in preservation can occur within the same skeletal population, depending on gender or age – female and juvenile skeletons seem to be frailer and tend to be destroyed more often than the male and adult skeletons (Masset 1973; Dutour 1989; Bello et al. 2006).

The palaeodemographic structure of the osteoarchaeological series needs to be known in palaeoepidemiology, especially when studying the prevalence of diseases having a gender- or age-specific prevalence rate, as is the case for some infectious diseases.

Preservation also depends on anatomical localisation; some parts of the skeleton (hand and foot, ribs, spine) are more delicate and, consequently, more often missing than other parts (Fig. 8.2). This differential preservation must be compared with the skeletal distribution of the diseases studied, taking into account the preferential localisation of a given infectious disease. For example, as the osseous involvement of TB frequently concerns the spine and extremities, it is of interest to know something about the preservation of these skeletal elements in the series. For leprosy, information about the preservation state of the facial skeleton (especially the nasal aperture and palate areas) and hand and foot bones is necessary to evaluate the material on which the calculation of prevalence was made. For treponematosis, although preferentially localised to a more robust part of the skeleton, the evaluation of prevalence must take into account the preservation state of tibiae and skulls.

8.4.4 Crude and Corrected Prevalences

Few methods that take taphonomy into account in calculating past prevalences have been developed. In most studies, estimations correspond to the ratio of the number of cases / total number of skeletons – termed "crude prevalence" by Waldron (1994). This measure of prevalence does not really consider the preservation state of the osseous remains. As mentioned above, each skeletal collection has its own preservation profile, and this crude method thus weakens the validity of comparative work. If we consider a theoretical skeletal collection of 200 individuals to evaluate the prevalence of tuberculosis, the identification of three cases of Pott's disease in this sample gives a prevalence rate of 1.5%. Taking into consideration the fact that 50 spines are almost totally missing, and that 50 others are too incomplete or fragmentary to yield valuable observations, the prevalence corresponds in fact to three observations on 100 spines, i.e. 3%. We recommend a correction of the prevalence using the formula: corrected prevalence, $C_rP = n/N{-}a$ (where a is the number of bones affected by disease that are *not* observable). We call this method correction by representation. Reducing the denominator increases the prevalence rate. Waldron suggests that prevalence in the missing parts can be assumed to be proportional to that

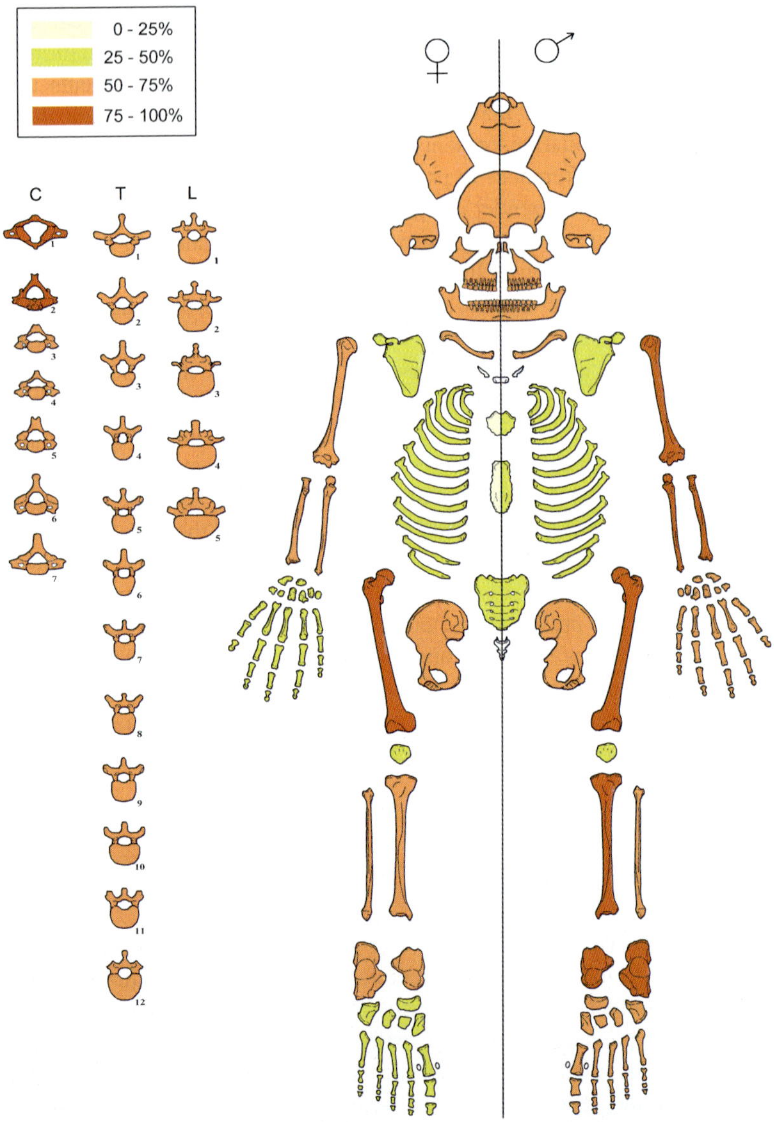

Fig. 8.2 Variation of skeletal preservation in function of the localisation and gender in a given skeletal collection (Bello 2000; Bello et al. 2006)

in the preserved parts, which validates C_rP for the totality of the sample. He also proposes considering the crude prevalence n/N (including *a*) as a minimal rate (*none* of the missing spines was affected) and a maximal prevalence rate of n+a/N (*all* of the missing spines were affected), the true rate lying somewhere in between. We propose another possibility to re-evaluate the crude ratio, which is to counterbalance the crude prevalence by a factor, *F*. The idea is to take into account the number of observable skeletal elements (vertebrae in this case), which are essential for the

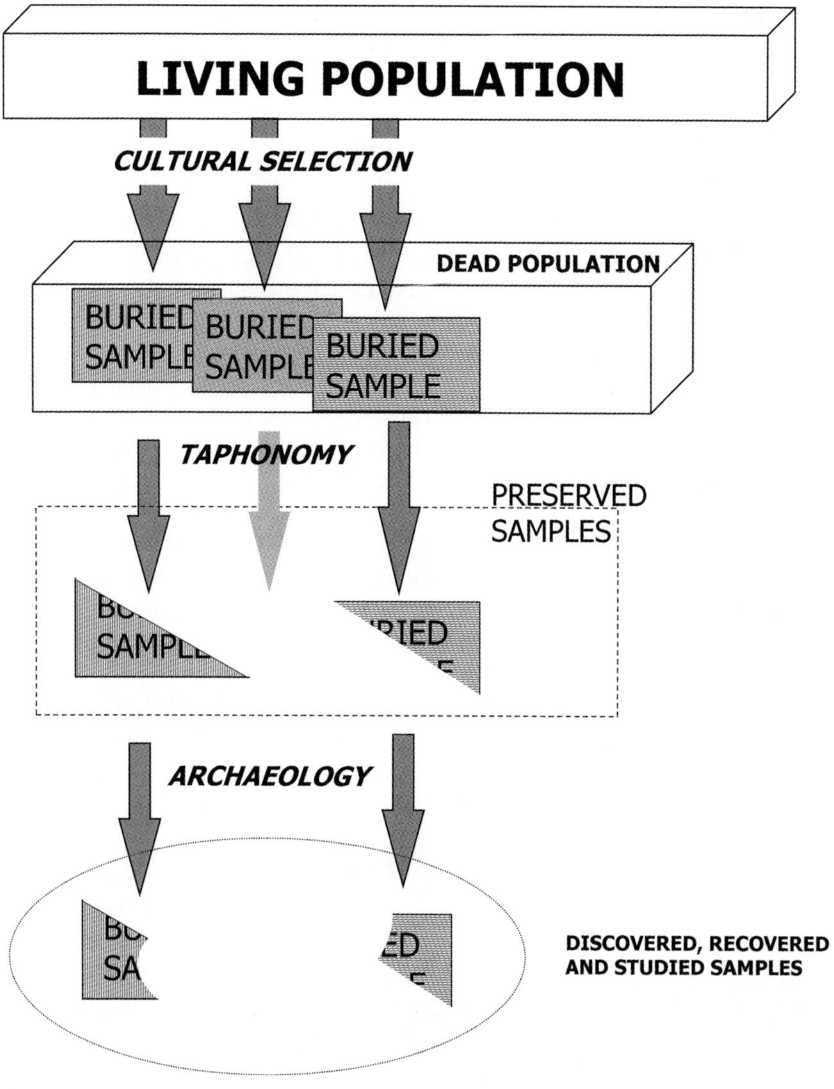

Fig. 8.3 Scheme of the three steps (cultural, taphonomic, archaeological) going from an ancient living population to its remains

diagnosis of the disease – TB in our example (Dutour et al. 2003). The ratio of the theoretical effective number / observable effective number represents the factor *F*. To take a real example, on the Hungarian collection from Balcsamas dating from the seventeenth century, Maczel (2003) found evidence of 15 cases of vertebral TB, giving a crude prevalence rate of 14%. However, among the 2,675 possible vertebrae (theoretical number), only 2,408 were represented. Thus, our counterbalance factor *F* is equal to 1.11. The counterbalanced prevalence is thus 15.5%.

The purpose of the above is to provide a method to calculate prevalence adapted to each sample that will be valid for comparative studies. However, according to

Bello et al. (2006) reliable results can be obtained if one compares prevalence in series showing a similar preservation pattern.

8.4.5 *Archaeology and Related Studies*

The constitution of a skeletal series depends mainly on archaeology (Fig. 8.3). Frequently, only part of a cemetery has been discovered, and/or excavations may have been carried out on no more than a segment of the unearthed part of the cemetery. The recovery may then concern only part of the excavated area, or only parts of the skeletons (e.g. skull and long bones, which were, until recently, considered the most informative elements for anthropologists). Anthropological study following excavation may be limited (e.g. to sex and age distribution only), and the storage of these skeletal series can make them difficult to study in their totality by palaeoepidemiologists. Hence we see the implications of other parameters of sample selection. The ideal case would be the excavation of a site in its totality, without any selection in the recovery of the osteological sample, with appropriate storage, contributing to open skeletal libraries.

We can thus appreciate the challenge more clearly: palaeoepidemiology mainly concerns the study of diseases in different skeletal samples, the latter having been to a greater or lesser degree selected from past populations by different factors in quite variable and unknown proportions.

Another important point concerning the sample structure, called intrinsic factor by Waldron (1994), in the characterisation of the nature of the sample should be highlighted.

8.5 Intrinsic Factor: Structure of the Sample

Theoretically, human skeletal remains are representative of a variable part of a dead population, which itself derives from the living population. This constitutes a major distinction from epidemiology, as palaeoepidemiologists study diseases in a community of the dead. According to Waldron (1994) 'it is surprising how often this fact is overlooked'.

The dead population differs in sex and age distribution from the living population. In less developed parts of the world – which we presume more closely resemble the past than more developed societies – the demographic structure of the mortality curve is the reverse of that for the living population. Over 40% of the living population is 15 years old or younger. The mortality curve, on the other hand, shows high mortality of the 0- to 5-year-old cohort, relative stability between the ages of 5 and 35, a progressive increase to age 55, and a dramatic increase after age 55, i.e. a typical U-shaped profile (Fig. 8.4).

Ideally, the entire dead part of a given population would be preserved in a single cemetery, enabling us to reconstruct the structure of the living population from the age distribution of the skeletal population. However, the relationships between the

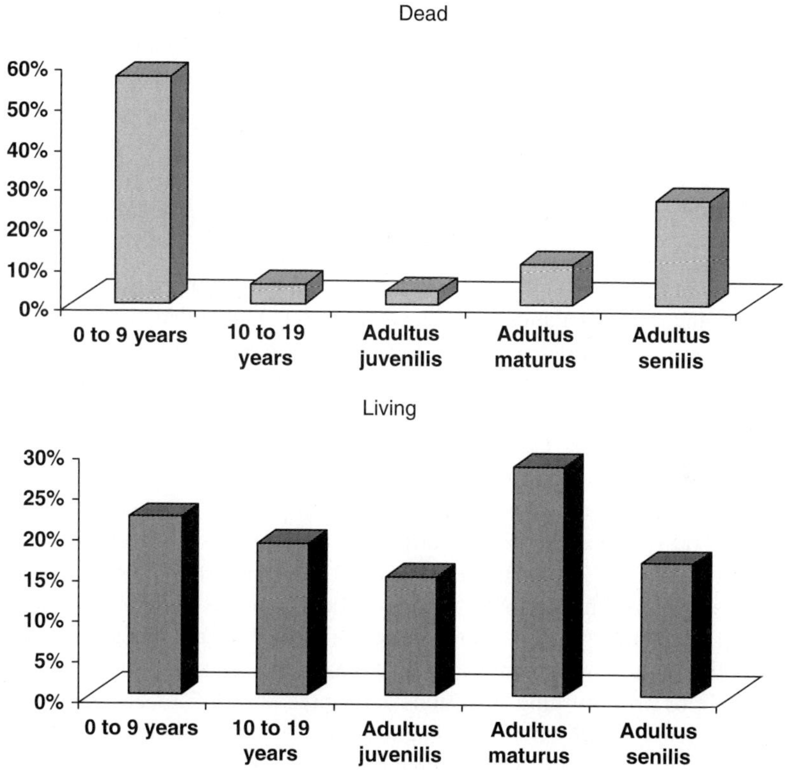

Fig. 8.4 Distribution of age categories for mortality and living profiles (pooled data from several historical pre-Jennerian populations and from the present populations of several undeveloped countries)

two curves depend on other parameters as well. An improved economy, for example, will modify the demographic pattern, making it older and reducing the mortality of its youngest members. Although the application of the demographic pattern of an undeveloped society to past populations is probably adequate in many cases, we must nonetheless consider other patterns, especially those in developing countries.

The ideal situation in palaeoepidemiology is encountered when the dead population has the same structure as the living one, for example when a non-selected part of the population (or the population in its entirety) suddenly disappears. This is what we called the 'Pompeii model' (Dutour et al. 1998, 2003). Working on other types of material corresponding to massive death occurring over a short time period, with no biological or cultural selection, constituted by skeletal series coming from plague epidemics, we assumed (Dutour et al. 1994) and subsequently demonstrated (Dutour et al. 1998, 2003) that such samples correspond to the criteria of the 'Pompeii model' and are very well suited to palaeoepidemiological analyses. Indeed, any peculiarities exhibited can minimise or even cancel out some of the common extrinsic or intrinsic biases observed in skeletal collections. Thus, such series can provide a more accurate picture of the palaeoepidemiological situation of some diseases than can be observed in more common types of material. The palaeoepidemiology of tuberculosis can be considered as an example.

8.6 Palaeoepidemiology of Tuberculosis

Tuberculosis is a good example of a re-emerging disease. Its prevalence is once more on the rise, and recent statistics place its mortality rate higher than that of AIDS. TB might become a major problem, especially if we take into account antibiotic-resistant germs, which are on the increase, and its future may very well be similar to its past. What we know of its past is limited mainly to mortality records of the nineteenth and early twentieth centuries. The classical data are that TB infection increased in the nineteenth century, its spread facilitated by urbanisation and overcrowding. In late nineteenth century France, the mortality rate from phthisis was between 3.08 and 3.69 per 1,000 (Bello et al. 1999); during the same period in Germany, mortality from TB was 2.6 per 1,000 (Alfer 1892, quoted in Ortner 2003). Our knowledge of the situation prior to this period is very poor, being limited to some rare historical records of mortality, such as the London Bills of Mortality beginning in the seventeenth century, which indicated that death by "consumption" (pulmonary tuberculosis or primary lung infection) accounted for 20% of all deaths during non-plague years (Clarkson 1975). The accuracy of diagnosis in the seventeenth century, however, was poor. A more reliable gauge is skeletal populations.

Since there is considerable uncertainty concerning the assignation of TB as the causative agent of the macro-morphological bone changes on which detection of TB infection has been mainly based in osteoarchaeological material (Waldron 1999), attention has focussed on the molecular level in search of a more reliable diagnosis and, consequently, more reliable disease frequencies in past populations. As a consequence, molecular biological techniques developed during the last decade have greatly broadened the diagnostic horizon in palaeopathology, not only by confirming the macroscopical diagnosis as a result of providing direct, demonstrative proof of tuberculous infection, but also by helping to identify new criteria for differential diagnosis.

Morphological techniques often do not allow the recognition of TB lesions, and the more specific identification of the disease agents is even more difficult, since human- and bovine-hosted TB, the two main human-affecting members of the *Mycobacterium tuberculosis* complex (MTC), produce anatomically similar bone changes (Ortner 1999). However, despite the fact that members of the MTC share many common characteristics, they, as well as other *Mycobacteria*, can be differentiated on the biomolecular level. The biomolecular analysis of archaeological human remains for TB has proved to be efficient. Such studies have been conducted in mummies (Salo et al. 1994; Nerlich et al. 1997; Crubézy et al. 1998; Pap et al. 1999; Zink et al. 2001), bone remains (Spigelman and Lemma 1993; Baron et al. 1996; Taylor et al. 1996, 1999; Fearman et al. 1999; Dutour et al. 1999; Haas et al. 2000) and even in calcified tissues (Donoghue et al. 1998; Pálfi et al. 1999), proving that fragments of ancient mycobacterial DNA can survive for long periods, probably due to their tough cell wall, and can provide direct evidence of TB infection. Such studies furnished evidence, from distinct genetic loci, for the presence of DNA fragments from Mycobacteria (65 kDa antigen gene) and more specifically from organisms belonging to the MTC (IS6110, rpoB) (Taylor et al. 1999; Haas et al. 2000; Mays et al. 2001). With the help of such biomolecular analyses, more reliable diagnosis of both typical and atypical morphological alterations can be developed, thus determining new diagnostic criteria involving more minor changes such as vertebral hypervascularisation (Ménard 1888; Baker 1999), rib periostitis (Kelley and Micozzi 1984; Roberts et al. 1994), and endocranial changes (Schultz 1999; Hershkovitz et al. 2002). An important source of tuberculous alterations can be found in anatomical collections where the cause of death is recorded. The search for new diagnostic criteria was extended to the United States [the Hamann-Todd (Kelley and Micozzi 1984) and Terry Collections (Roberts et al. 1994)] as well as to Portugal, where *Mycobacterium tuberculosis* infection in the Coimbra Identified Skeletal Collection was confirmed by the use of biomarkers (Santos and Roberts 2001).

However, mycobacterial DNA can be detected even in bones without morphological changes (Fearman et al. 1999; Zink et al. 2001). This point leads to the question of infection versus exposure to infection, which is especially relevant in molecular palaeo-epidemiology. The question of the meaning of negative or positive molecular results is still broadly open from a palaeoepidemiological point of view: a negative result can signify either lack of infection or a molecular taphonomical problem; a positive result, with the exception of contamination, is not necessarily related to disease, as it can also provide testimony of exposure to the infection.

The prevalence of skeletal lesions can thus help evaluate the frequency of the disease versus exposure to the disease; the latter is thought to be very high, even generalised, if TB infection was present in a given ancient population.

We attempted to estimate the minimal prevalence of TB on plague skeletal material and other material studied. As mentioned above, we encountered several methodological problems:

1. Reconstruction of global TB prevalence from skeletal lesions depends on the frequency of TB skeletal involvement over the total number of TB infections. According to the literature (Nathanson and Cohen 1941; Lafond 1958; Kelley and Micozzi 1984; Davies et al. 1984; Aufderheide and Rodriguez-Martin 1998; Ortner 2003), this varies from 3% to 9%. Hence, to avoid overestimation, we can assume a minimal prevalence from the minimal frequency of skeletal involvement.
2. TB skeletal infection mainly affects the youngest part of the population (Sorrel and Sorrel-Dejerine 1932), with more than 60% of all victims being under the age of 20. Poor preservation or absence of the youngest individuals in an osteo-archaeological series will underestimate the reconstructed frequency of skeletal TB.
3. Bone repartition of TB involvement mainly affects the spine (25–50% of skeletal TB cases; Steinbock 1976) and the extremities. These parts of the skeleton are, unfortunately, often poorly preserved.
4. A palaeopathological diagnosis is established on morphological (osteological and radiological) criteria, defined by comparison of clinical, radiological, and pathological records (Sorrel and Sorrel-Dejerine 1932). The significant TB

> prevalence in the past, as evidenced by documentary sources (Cronje 1984) contrasts with the paucity of palaeopathological evidence (Stirland and Waldron 1990). It has been suggested that the usual palaeopathological diagnostic criteria for skeletal tuberculosis are inadequate (Roberts et al. 1994). Biomolecular analysis of *Mycobacterium tuberculosis* DNA in presumed palaeopathological cases may confirm the diagnosis (Spigelman and Lemma 1993; Baron et al. 1996; Taylor et al. 1996, Dutour et al. 1999, Pàlfi et al. 1999, Salo et al. 1994, Taylor et al. 1996, Nerlich et al. 1997, Crubézy et al. 1998, Haas et al. 2000, Zink et al. 2001, Mays et al. 2001).

If we take a rough look at some large osteoarchaeological collections (numbering a total of 5,848 skeletons) from Hungary, dating from the seventh to the seventeenth centuries, a reconstruction from skeletal lesions of the minimal prevalence of TB infection in the population showed variations depending on chronology (Pàlfi and Marcsik 1999): between the seventh and eighth centuries (the Avar era), it represents 23% (crude prevalence rate: 0.7); during the tenth century Hungarian conquest, 0% (but some cases of leprosy have been described; Pàlfi 1991); between the eleventh and thirteenth centuries, 8.6% (crude prevalence rate: 0.26); and for the period from the fourteenth to the seventeenth centuries, 31% (crude prevalence rate: 0.95). This osteoarchaeological material does not provide us with the desired criteria for palaeoepidemiology, i.e. short periods of time, absence of selection, Pompeii-like palaeodemographic structure.

On our plague material, for one of our series (L'Observance), diagnosis was established both morphologically and molecularly on 3 individuals out of the 179 that can be observed (crude prevalence rate: 1.67%). These three samples also gave positive

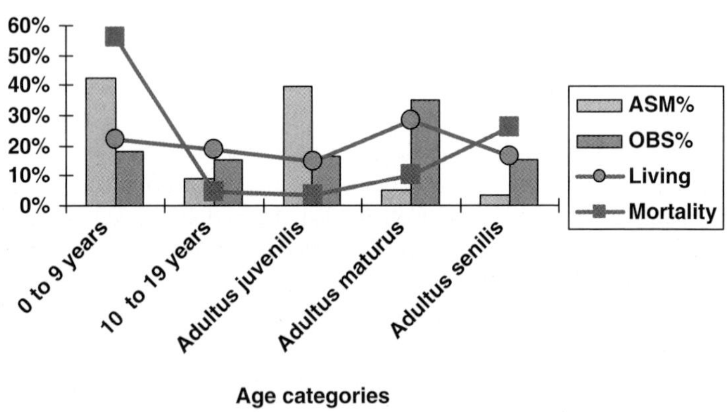

Fig. 8.5 Differences in sample structure. A sample from a plague mass grave (OBS) dating from the eighteenth century is much more similar to the age category distribution of a contemporaneous living population (historical demographic data). Sample ASM (slave cemetery, from the same period) exhibits clear differences in its structure, even in comparison with a mortality profile (same data): over-representation of young adults is obvious. This difference in structure is sufficient to explain the results obtained by reconstructing minimal prevalences for tuberculosis in these two samples, as well as the poorly realistic result for ASM (55% and over 100%, respectively)

molecular results (Zink et al. 2001); other 'control' samples with no lesions remained negative. In the same manner, if we base our estimation on 3% of skeletal involvement, the minimal prevalence of TB infection in the population in 1722 was about 55%. This prevalence seems to be very high for eighteenth century material; however, we should bear in mind the frequency of tubercular infection observed in undeveloped countries 40 years ago, e.g. 37% in Phnom-Penh in 1966 (Nguyen 1988).

Moreover, a study of a contemporaneous eighteenth–nineteenth century slave cemetery in the French West Indies (Courtaud et al. 2005) revealed six cases of vertebral tuberculosis on 148 preserved spines (crude prevalence: 4%). The minimal prevalence of TB is over 100%, suggesting a generalised TB infection in this population of slaves or, more likely, an effect of sample structure in the cemetery population, due to the addition of successive dead parts during a period with high TB prevalence (over about one century) in the living population of slaves (Fig. 8.5).

Such results highlight the need for reliable palaeopathological material for palaeoepidemiology, especially in the reconstruction of past infectious diseases.

References

Aufderheide A, Rodriguez-Martin C (1998) The Cambridge encyclopedia of human paleopathology. Cambridge University Press

Baker BJ (1999) Early manifestations of tuberculosis in the skeleton. In: Pàlfi Gy, Dutour O, Deak J, Hutas I (eds) Tuberculosis past and present. Golden Book, Budapest-Szeged, pp 301–310

Baron H, Hummel S, Herrmann B (1996) *Mycobacterium tuberculosis* complex DNA in ancient human bones. J Archaeol Sci 23:667–671

Bello S (2000) Taphonomie des restes osseux humains. Effet des processus de conservation du squelette sur les paramètres anthropologiques. PhD Université de la Méditerranée, Faculté de Médecine de Marseille Universita' degli studi di Firenze

Bello S, Signoli M, Maczel M, Dutour O (1999) Evolution of mortality due to tuberculosis in France (18th–19th). In: Pàlfi Gy, Dutour O, Deak J, Hutas I (eds) Tuberculosis past and present. Golden Book, Budapest-Szeged, pp 93–104

Bello S, Thomann A, Signoli M, Dutour O, Andrews P (2006) Age and sex bias in the reconstruction of past population structures. Am J Phys Anthropol 129:24–38

Blaizot F, Alix G, Ferber E (2003) Le traitement funéraire des enfants décédés avant un an dans l'Antiquité: études de cas. Bull Mem Soc Anthropol Paris 15:1–2

Boldsen JL (2001) Epidemiological approach to the paleopathological diagnosis of leprosy. Am J Phys Anthropol 115:380–387

Brothwell D (1961) The palaeopathology of early man: an essay on the problems of diagnosis and analysis. J R Anthropol Inst 91:318–344

Clarkson L (1975) Death, disease and famine in pre-industrial England. Gill/Macmillan, Dublin

Cohen MN, Crane-Kramer G (2003) The state and future of paleoepidemiology. In: Greenblatt C, Spigelman M (eds) Emerging pathogens. Oxford University Press, Oxford, pp 79–91

Combes C (2001) Parasitism ecology and evolution of intimate interactions. Chicago University Press, Chicago, IL

Courtaud P, Romon T, Amouric H, Bruzek J, Dutour O, Tatilon C, Vallauri L, Verrand L (2005) Les cimetières d'esclaves. Exemple de l'ensemble d'Anse Sainte-Marguerite (le Moule, Guadeloupe). In: Ardagna Y, Boëtsch G, Dutour O, Lalys L, Signoli M (eds) L'homme et ses images. UMR 6578 Marseille, pp 325–338

Cronje G (1984) Tuberculosis mortality decline in England and Wales, 1851–1910. In: Woods R, Woodward J (eds) Urban disease and mortality in 19th century England. Batsford, London, pp 79–101

Crubézy LB, Poyeda JD, Clayton J, Crouau-Roy B, Montagnon D (1998) Identification of Mycobacterium DNA in an Egyptian Pott's disease of 5,400 years old. C R Acad Sci III 321:941–951

Darwin C (1859) The origin of species. Murray, London

Davies PDO, Humphries MJ, Byfield SP, Nunn AJ, Darbyshire JH, Citron ICM, Fox W (1984) Bone and joint tuberculosis. J Bone Joint Surg 66B:320–330

Delpech JM (1816) Précis élémentaire des maladies réputées chirurgicales. Méquignon-Marvis, Paris

Donoghue HD, Spigelman M, Zias J, Gernaey-Child AM, Minnikin DE (1998) *Mycobacterium tuberculosis* complex DNA in calcified pleura from remains 1400 years old. Lett Appl Microbiol 27:265–269

Drancourt M, Raoult D (2005) Palaeomicrobiology: current issues and perspectives. Nat Rev Microbiol 3:23–35

Dutour O (1989) Hommes fossiles du Sahara. CNRS, Paris

Dutour O, Signoli M, Georgeon E, da Silva J (1994) Le charnier de la Grande peste de Marseille (rue Leca): données de la fouille de la partie centrale et premiers résultats anthropologiques. Préhist Anthropol Médit 3:191–203

Dutour O, Signoli M, Pàlfi G (1998) How to reconstruct the epidemiology of infectious diseases in the past? In: Greenblatt C (ed) Digging for pathogens. Balaban, Rehovot, pp 241–263

Dutour O, Pàlfi Gy, Brun JP, Bérato J, Panuel M, Haas CJ, Zink A, Nerlich AG (1999) Morphological, paleoradiological and paleomicrobiological study of a French medieval case of Tuberculosis spondylitis with cold abscess. In: Pàlfi Gy, Dutour O, Deak J, Hutas I (eds) Tuberculosis past and present. Golden Book, Budapest-Szeged, pp 395–402

Dutour O, Ardagna Y, Maczel M, Signoli M (2003) Epidemiology of infectious diseases in the past: Yersin, Koch and the skeletons. In: Greenblatt C, Spigelman M (eds) Emerging pathogens. Oxford University Press, Oxford, pp 151–165

Fearman M, Jankauskas R, Gorski A, Bercovier H, Greenblatt CL (1999) Detecting *Mycobaterium tuberculosis* DNA in medieval skeletal remains from Lithuania: In: Pàlfi Gy, Dutour O, Deak J, Hutas I (eds) Tuberculosis past and present. Golden Book, Budapest-Szeged, pp 371–378

Gerstman BB (2003) Epidemiology kept simple: an introduction to traditional and modern epidemiology, 2nd edn. Wiley-Liss, Hoboken, NJ

Haas CJ, Zink A, Molnár E, Szeimies U, Reischl U, Marcsik A, Ardagna Y, Dutour O, Pálfi GY, Nerlich AG (2000) Molecular evidence for different stages of tuberculosis in ancient bone samples from Hungary. Am J Phys Anthropol 113:293–304

Hershkovitz I, Greenwald CM, Latimer B, Jellema LM, Wish-Baratz S, Eshed V, Dutour O, Rothschild BM (2002) Serpens endocrania symmetrica (SES): a new term and a possible clue for identifying intrathoracic disease in skeletal populations. Am J Phys Anthropol 118:201–216

Kelley MA, Micozzi MS (1984) Rib lesions in chronic pulmonary tuberculosis. Am J Phys Anthropol 65:381–386

Lafond EM (1958) An analysis of adult skeletal tuberculosis. J Bone Joint Surg 40A:346–364

Maczel M (2003) Sur les traces de la tuberculose: critères diagnostiques des atteintes infectieuses du squelette humain. PhD Université de la Méditerranée, Faculté de Médecine de Marseille Szeged University

Maczel M, Dutour O, Ardagna Y, Panuel M, Zink A, Nerlich AG, Pàlfi G (2005) La série médiévale de l'Abbaye de la Celle, Var: données anthropologiques et paléopathologiques. In: Ardagna Y (ed) L'homme et ses images. UMR, Marseille 6578:393–402

Masset C (1973) Influence du sexe et de l'âge sur la conservation des os humains. In: L'Homme, hier et aujourd'hui. Recueil d'études en Hommage à André Leroi-Gouhan. Cujas, Paris, pp 333–343

Mays S (1992) Taphonomy factors in a human skeletal assemblage. Circaea 9:54–58

Mays S, Taylor GM, Legge AJ, Young DB, Turner-Walker G (2001) Paleopathological and biomolecular study of tuberculosis in a medieval skeletal collection from England. Am J Phys Anthropol 114:298–311

Medilexicon (2007) Medical Dictionary. http://wwwmedilexiconcom/© MediLexicon International

Menard V (1888) Tuberculose vertébrale. Asselin and Houzeau, Paris

Nathanson E, Cohen W (1941) A statistical and roentgen analysis of two hundred cases of bone and joint tuberculosis. Radiology 36:550–567

Nerlich AG, Haas CJ, Zink A, Szeimies U, Hagedorn HG (1997) Molecular evidence for tuberculosis in an ancient Egyptian mummy. Lancet 349:1760–1763

Nguyen TA (1988) Epidémiologie tropicale. ACCT

Ortner DJ (1999) Paleopathology: implications for the history and evolution of tuberculosis. In: Pàlfi Gy, Dutour O, Deak J, Hutas I (eds) Tuberculosis past and present. Golden Book, Budapest-Szeged, pp 255–262

Ortner DJ (2003) Identification of pathological conditions in human skeletal remains. Academic, Amsterdam

Pàlfi G (1991) The first osteoarchaeological evidence of leprosy in Hungary. Int J Osteoarchaeol 1:99–102

Pàlfi G, Marcsik A (1999) Palaeoepidemiological data of tuberculosis in Hungary. In: Pàlfi Gy, Dutour O, Deak J, Hutas I (eds) Tuberculosis past and present. Golden Book, Budapest-Szeged, pp 533–542

Pàlfi G, Ardagna Y, Molnar E, Dutour O, Panuel M, Haas CJ, Zink A, Nerlich AG (1999) Coexistence of tuberculosis and ankylosing spondylitis in a 7–8th century specimen. In: Pàlfi Gy, Dutour O, Deak J, Hutas I (eds) Tuberculosis past and present. Golden Book, Budapest-Szeged, pp 403–413

Pap I, Jozsa L, Repa I, Bajzik G, Lakhani SR, Donoghue HD, Spigelman M (1999) 18–19th century tuberculosis in natural mummified individual (Vacs, Hungary). In: Pàlfi Gy, Dutour O, Deak J, Hutas I (eds) Tuberculosis past and present. Golden Book, Budapest-Szeged, pp 421–428

Roberts C, Lucy D, Manchester K (1994) Inflammatory lesions of ribs: an analysis of the Terry Collection. Am J Phys Anthropol 95:169–182

Salo WL, Aufderheide AC, Buikstra J, Holcomb TA (1994) Identification of *Mycobacterium tuberculosis* in a pre-Columbian Peruvian mummy. Proc Natl Acad Sci USA 91:2091–2094

Santos AL, Roberts CA (2001) A picture of tuberculosis in young Portuguese people in the early 20th century: a multidisciplinary study of the skeletal and historical evidence. Am J Phys Anthropol 115:38–49

Santos AL, Roberts CA (2006) Anatomy of a serial killer: differential diagnosis of tuberculosis based on rib lesions of adult individuals from the Coimbra Identified Skeletal Collection. Am J Phys Anthropol 130:38–49

Schultz M (1999) The role of tuberculosis in infancy and childhood in prehistoric and historic population. In: Pàlfi Gy, Dutour O, Deak J, Hutas I (eds) Tuberculosis past and present. Golden Book, Budapest-Szeged, pp 503–510

Sellier P (1996) La mise en évidence d'anomalies démographiques et leur interprétation: population, recrutement et pratiques funéraires du tumulus de Courtesoult. In: Piningre J-F (ed) Nécropoles et société au premier âge du Fer: le tumulus de Courtesoult (Haute-Saône). Documents d'Archéologie Française 54:188–202

Sorrel E, Sorrel-Dejerine C (1932) Tuberculose osseuse et ostéo-articulaire. Masson, Paris

Spigelman M, Lemma E (1993) The use of the polymerase chain reaction (PCR) to detect *Mycobacterium tuberculosis* in ancient skeletons. Int J Osteoarcheol 3:137–343

Steinbock RT (1976) Paleopathological diagnosis and interpretation. Thomas, Springfield, IL

Stirland A, Waldron T (1990) The earliest cases of tuberculosis in Britain. J Archaeolog Sci 17:221–230

Taylor GM, Crossey M, Saldhana J, Waldron T (1996) DNA from *Mycobacterium tuberculosis* identified in mediaeval human skeletal remains using polymerase chain reaction. J Archaeol Sci 23:789–798

Taylor GM, Goyal M, Legge AJ, Shaw RJ, Young D (1999) Genotypic analysis of *Mycobacterium tuberculosis* from mediaeval human remains. Microbiology 145:899–904

Ubelaker D (1998) Ancient disease in anthropological context. In: Greenblatt C (ed) Digging for pathogens. Balaban, Rehovot, pp 175–198

Van Valen L (1973) A new evolutionary law. Evol Theory 1:1–30

Wood JW, Milner GR, Harpending HC, Weiss KM (1992) The osteological paradox. Problems of inferring prehistoric health from skeletal samples. Curr Anthropol 33:343–370

Waldron T (1985) DISH at Merton Priory: evidence for a new occupational disease? BMJ 291:1762–1763

Waldron T (1994) Counting the dead: the epidemiology of skeletal populations. Wiley, Chichester

Waldron T (1999) The paleoepidemiology of tuberculosis: some problems considered. In: Pàlfi Gy, Dutour O, Deak J, Hutas I (eds) Tuberculosis past and present. Golden Book, Budapest-Szeged, pp 471–478

Watts DJ (1989) Infant burials and Romano–British christianity. Archaeol J 146:372–383

Zink A, Haas C, Reischl U, Szeimes U, Nerlich AG (2001) Molecular analysis of skeletal tuberculosis in an ancient Egyptian population. J Med Microbiol 50:355–366

Chapter 9
Past Plague

Michel Drancourt and Didier Raoult (✉)

Abstract The recent discovery, by two independent teams, of *Yersinia pestis* DNA in human remains dating from two historical plague pandemics, has generated renewed interest in the epidemiology of past plague epidemics. A scenario involving one of the three different *Y. pestis* pathovars identified at the time in each of the three pandemics was proposed in 1951. Palaeomicrobiologic and genetic data support an alternative scenario, with an Orientalis-like strain originating from Asia being responsible for all three plague pandemics.

9.1 Introduction

Plague caused by *Yersinia pestis* has been responsible for millions of deaths for at least two millennia (Perry and Fetherston 1997). In recent times, renewed interest in plague has been generated due to the emergence of multi-resistant strains of *Y. pestis* (Chanteau et al. 1998; Galimand et al. 1997) and the growing recognition of the potential of *Y. pestis* as an agent of biological warfare (Inglesby et al. 2000) (http://www.bt.cdc.gov).

Plague, a zoonose, primarily affects rodents. Man, an incidental host, is infected by rat fleas (Perry and Fetherston 1997). The flea acquires *Y. pestis* from the blood of a bacteraemic reservoir animal. The infection is restricted to the gastrointestinal tract of the flea. Typically, plague is thought to exist indefinitely in rodent populations in so-called enzootic (maintenance) cycles that involve transmission between partially resistant rodents (enzootic or maintenance hosts) and their fleas. Not infrequently, the disease spreads from enzootic to more susceptible animals (epizootic or amplifying hosts), causing rapidly spreading die-offs (epizootics) (Gage et al. 1995). Anthropophilic rodent fleas may transmit *Y. pestis* to humans and are believed to be responsible for human plague. Following the third pandemic of

Didier Raoult
Unité des Rickettsies, CNRS UMR 6020, Faculté de Médecine, Université de la Méditerranée, 27 Boulevard Jean Moulin, 13385 Marseille Cedex 5, France
E-mail: Didier.Raoult@medecine.univ-mrs.fr

D. Raoult and M. Drancourt (eds.), *Paleomicrobiology: Past Human Infections.*

plague at the end of the nineteenth century, Yersin (1894), Simond (1898) and Raybaud (Gauthier and Raybaud 1903) established that rat fleas transmitted *Y. pestis* to human subjects. This was later endorsed by the then Indian National Advisory Committee on plague (1907). Transmission by human ectoparasites, including the body louse *Pediculus humanus corporis*, has also been observed during twentieth century epidemics (Drancourt et al. 2006). Recent experimental data confirmed the potential role of body lice as a vector of plague (Houhamdi et al. 2006). This fact may explain the particular epidemiological patterns that emerged during large historical pandemics (Drancourt et al. 2006)

Based on the description of outbreaks associated with bubonic lesions, history can trace three great pandemics of plague (Perry and Fetherston 1997). We know no other cause of bubonic fever outbreak, and it has been speculated that these pandemics were caused by the same organism. The first great pandemic was that known by historians as the Justinian plague (541–544 A.D.). A detailed account of the outbreak of plague in Constantinople (modern Istanbul, Turkey) was given by Procopius of Caesarea in his book "De bello persico" (Procopius 1914). According to this source, the outbreak involved bubonic plague and had killed an estimated 50% of the inhabitants of the Byzantine Empire by the year A.D. 565. Resurgence of this pandemic was noted every 8–12 years until the eighth century. In 1347, plague re-emerged and this second pandemic claimed an estimated 17–28 million human lives (Ziegler 1991). It was described properly as an outbreak of bubonic fever by the fourteenth century physician and plague witness Guy de Chaulliac in 1363 (Enselme 1969). Resurgences were observed until the end of the eighteenth century. In France, the last resurgence of this second pandemic occurred in Marseilles and claimed 40–50% deaths among the 100,000 inhabitants (Signoli et al. 1996). The third pandemic (1855) established stable enzootic foci in every continent except Antarctica. The infectious nature of plague was not understood until Alexandre Yersin cultured Gram-negative bacilli from enlarged lymph node aspirate obtained from a case of bubonic plague during the Hong Kong epidemics in the mid-1890s (Yersin 1894). At around the same time, the Japanese bacteriologist Shibasaburo Kitasato independently announced the isolation of the plague bacillus. However, the initial description of the microorganism he isolated might have included a contaminating pneumococcus (Kitasato 1894). The fact that Yersin inoculated high inoculum, buboe-derived material at room temperature instead of using an incubator may have been decisive in his success in isolation of the further named *Yersinia pestis*.

9.2 The Bacterium

Yersinia pestis is a non-motile, non-sporulating, Gram-negative biochemically unreactive member of the family *Enterobacteriaceae* of γ-proteobacteria. Although encapsulated, *Y. pestis* produces an envelope that contains the unique fraction 1 (Fr1) glycoprotein surface antigen. It dies rapidly if exposed to temperatures exceeding 40°C or desiccation. Three *Y. pestis* biotypes have been recognised on the basis of their abilities to convert nitrate to nitrite and to ferment glycerol (Perry and Fetherston 1997). The biotype Antiqua has both characteristics. The biotype Medievalis ferments glycerol but does not form nitrite. The biotype Orientalis forms nitrite but does not ferment glycerol. A fourth biotype, Microtus, has been proposed to accommodate Chinese isolates from *Microtus* sp. The latter biotype differs from Medievalis by its inability to ferment arabinose (Zhou et al. 2004). Russian authors developed an alternative classification scheme based on 18 phenotypic characters, which distinguished six subspecies found in the former Soviet Union with defined pathogenicity in guinea-pigs and geographic partition in former Soviet states (Anisimov et al. 2004) (Table 9.1).

Genetic analyses at the population level have indicated that *Y. pestis* diverged from its closely related *Yersinia pseudotuberculosis* ancestor [probably serotype O: 1b (41) 1,500–20,000 years ago (Achtman et al. 1999, 2004) along one branch (branch 0) supporting the human-avirulent Microtus isolate 91001 and pestoides isolates, and then diverged into two main branches: branch 1 comprising Orientalis and the African Antiqua isolates, and branch 2 comprising the Medievalis and the Asian Antiqua isolates (Achtman et al. 2004; Chain et al. 2006) (Fig. 9.1). These analyses, as well as single nucleotide polymorphism (SNP)-based analysis of complete genomes, therefore indicated that Antiqua is an inaccurate phylogenetic representation (Chain et al. 2004).

Isolates of the African Antiqua biotype are currently found in Central Africa, whereas those belonging to the Asian Antiqua biotype are found in south-eastern Russia, Manchuria, Mongolia and central and northern Asia; the biotype Medievalis is currently found around the Caspian Sea, Iranian Kurdistan and Southeastern Russia in Western Kazakhstan between the Volga and Ural rivers; the biotype Microtus is found in China and Tibet; and the biotype Orientalis is disseminated worldwide (Perry and Fetherston 1997) (Fig. 9.2).

Comparative genomics of five complete *Y. pestis* genomes including two Antiqua isolates (Chain et al. 2004) and one each of the Medievalis (Deng et al. 2002), Orientalis (Parkhill et al. 2001) and Microtus (Song et al. 2004) biotypes, along with the closely related *Y. pseudotuberculosis* genome (Chain et al. 2004), found unique features for each *Y. pestis* biotype (Table 9.2). The Orientalis biotype genome is unique in having not only a decreased number of coding sequences, but also a decreased number of predicted inactivated genes, a decreased number of RNA operons (6 copies of rRNA operons instead of 7; 70 tRNA genes instead of 72–73), and the accumulation of insertion sequences (44 copies of *IS* 100 insertion sequences instead of 30–75, 62 copies of *IS* 1541 instead of 43–67). It shares some of these characteristics in common with the African Antiqua strain. This evolution is typical of the bacteria causing human outbreaks as they have a higher rate of multiplication (Wren 2000).

Diversity among *Y. pestis* strains was first assessed by pulsed-field gel electrophoresis (PFGE). Analyses of a limited number of isolates have demonstrated that the "pulsotypes" were closely related to their corresponding biotypes (Lucier and Brubaker 1992; Rakin and Heesemann 1995). Strains of ribotype B were all of biovar Orientalis origin and were found over five continents, whereas those of ribotype

Table 9.1 Phenotypic typing of *Yersinia pestis* isolates adapted from Russian authors (Anisimov et al. 2004)

Y. pestis subspecies	Fermentation of Dependence on nutrition factors																					
	Rhamnose	Mellbiose	Arabinose	Glycerol	Melezitose	Nitrate reduction	Urease activity	Pesticin 1 production	Susceptibility to pesticin 1	Fibrinolytic activity	Coagulase activity	Leucine	Methionine	Arginine	Thiamine	Cysteine	Phenylalanine	Threonine	Tyrosine	Virulence for guinea pigs	Region of circulation	Biovar
pestis	−	−	+	−	−	+	±	+	−	+	+	±	±	−	−	±	±	+	−	+	Worldwide	Orientalis
pestis	−	−	+	+	−	+	−	+	−	+	+	±	+	−	−	±	±	+	−	+	Central Africa, central and northern Asia, China (Manchuria), Mongolia	Antiqua
causasica	+	+	+	+	−	+	−	−	+	−	−	±	+	+	+	±	+	+	NA	−	Transcaucasian highland, Mountain Dagestan	Antiqua
altaica	+	+	−	+	NA[a]	−	−	+	+	+	+	+	−	+	−	+	+	NA	NA	−	Mountain Altai	Medievalis
hissaria	+	+	−	+	±	−	±	+	±	+	+	+	+	−	−	+	+	−	−	−	Issarian Ridge	Medievalis
ulegeica	+	+	+	+	NA	−	−	+	±	+	+	−	−	−	−	+	+	NA	NA	−	Northeast Mongolia, Gobi Desert	Medievalis
talassica	+	+	−	+	−	−	+	+	−	+	+	+	NA	+	NA	+	+	NA	+	−	Talassian Ridge	Medievalis

[a] Not available

Fig. 9.1 A view of 1720 Marseilles' plague outbreak as painted by a contemporary witness, M. Serre

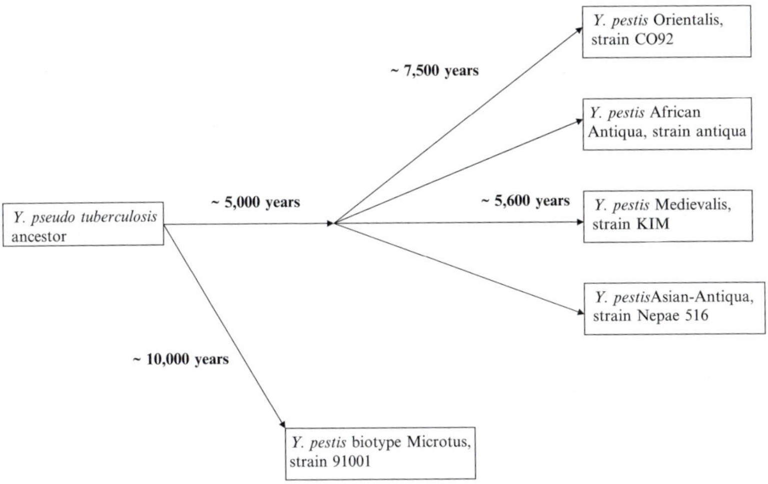

Fig. 9.2 Phylogenetic representation of the various biotypes of *Yersinia pestis* (adapted from Achtman et al. 2004 and Chain et al. 2006). The sequenced reference strain for each biotype is indicated

O were either biovar Medievalis or biotype Antiqua and were found in suspected remaining foci of the first and second pandemics (Guiyoule et al. 1994). However, individual isolates within one biotype demonstrated heterogeneity on PFGE. The latter suggests spontaneous rearrangements of DNA occurring in various isolates. Ribotyping subdivided 70 strains of *Y. pestis* into 16 ribotypes and found that two ribotypes (B and O) comprised the majority of isolates. The analysis of variable

Table 9.2 Comparative genomics of *Yersinia* spp. genomes

	Y. pestis					*Y. pseudotuberculosis*
Strain	CO92	Antiqua	KIM	Nepal 516	91001	IP32593
Biotype	Orientalis	African-Antiqua	Medievalis	Asian-Antiqua	Mictotus	/
Chromosome size (Mbp)	4653	47	46	453	4595	4.744
G+C content	47.64	47.7	47.64	47.58	47.65	47.61
Coding sequences	4,012	4,138	4,198	3,956	4,037	3,974
Average gene length (bp)	998	953	940	958	966	/
Predicted inactivated genes	36	85	86	49	98	NA
16S–23S–5S rRNAs	6	7	7	7	7[a]	7
Transfer RNAs	70	68	73	72	72	85
pMT size (bp)	96,210	96,471	100,990	100,918	106,642	/
pCDb size (bp)	70,305	70,299	70,504	NAa	70,159	68,526
pPCP size (bp)	9,612	10,777	10,961	10,778	9,609	/
Total IS elements	134	176	111	129	103	20
IS 100 elements	44	75	35	32	30	5
IS 285 elements	21	24	19	25	23	7
IS 1541 elements	62	67	49	64	43	5
IS 1661 elements	7	10	8	8	7	3
Reference	Le et al. 2001	Ingelsby et al. 2000	Kitasato 1894	Ingelsby et al. 2000	Lowell et al. 2005	Lucier and Brubaker 1992

[a]Not available

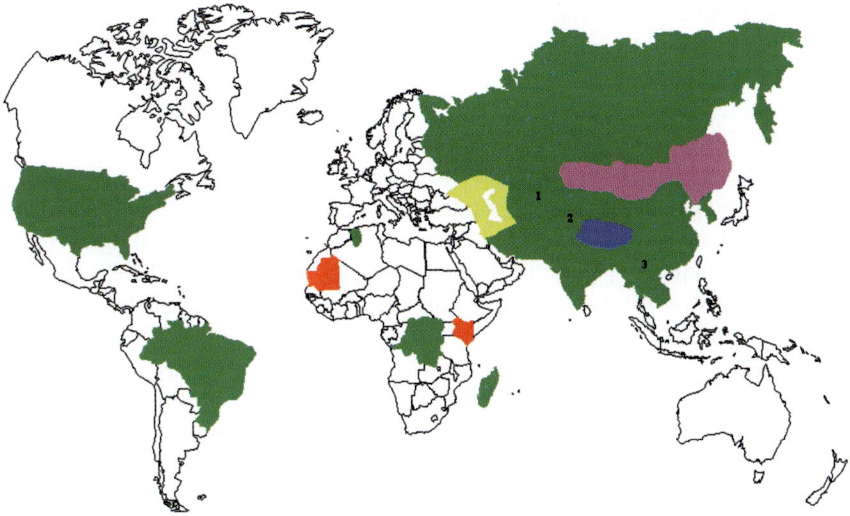

Fig. 9.3 Putative sources of historical plague pandemics (indicated by *numbers 1–3*) illustrates limited spread of Antiqua, Medievalis and Microtus biotypes and worldwide distribution of the Orientalis biotype. *Red* African Antiqua, *pink* Asian Antiqua, *yellow* Medievalis, *green* Orientalis, *blue* Microtus. This overall distribution was compiled from references Pollitzer 1954, Chanteau et al. 1998 and Inglesby et al. 2000

number tandem repeats (VNTR) on electrophoresis gels offered promising results for the typing of *Y. pestis* (Le et al. 2001). When applied to a large collection of 180 isolates originating from three continents, the analysis of 25 markers found 61 genotypes and distributed the three biotypes within three main branches (Pourcel et al. 2004). This method showed promise in the identification of sources of plague in modern cases of human plague (Lowell et al. 2005). However, VNTR analysis is restricted to cultured strains that yield high quality DNA.

Recently, we have developed a sequence-based method named multispacer sequence typing (MST) based on complete genome data by sequencing several intergenic spacers, some including tandem repeats and some SNPs (Drancourt et al. 2004). This has been shown to be effective in discriminating among biotypes in a large collection of *Y. pestis* isolates (Drancourt et al. 2004) (Fig. 9.3). Moreover, MST was successfully applied to the genotyping of uncultured strains from ancient, buried specimens (Drancourt et al. 1998). MST analyses single mutations in addition to tandem repeats – two molecular events with the same evolutionary significance.

9.3 Geographical Sources of Historical Plague Pandemics

The geographical origin of the first pandemic remains controversial. It is widely believed that plague was present in Ethiopia in 541 A.D. and spread quickly from Pelusium at the eastern limit of the Nile Delta, Egypt, through the Middle East to the Mediterranean basin. Although contemporary sources suggest either an Egyptian (Procopius 1914), or Ethiopian origin, from the kingdom of Axum, whether Eastern Africa was the ultimate source of the first pandemic remains doubtful. Other authors have suggested that first pandemic had its origins among wild gerbil populations in eastern Asia (McKeown 1988). Recent molecular evidence suggests that the first pandemic focus may have been Asia rather than Africa (Zhou et al. 2004).

The second pandemic probably originated in the steppes of Central Asia, where there was an epidemic of marmots. Chwolson, a Russian archaeologist, found inscriptions related to plague on memorial stones dating back to 1338–1339 in Nestorian graveyards near Issyk Kul Lake. However, it remains unclear how and when *Y. pestis* circulating in the marmot populations of the Middle and Central Asian mountains and Tibet, penetrated into the mountain savannas of Eastern and Central Africa. Ziegler (1991) and Pollitzer (1954) suggested that wild rodent populations in the area near Lake Issyk Koul in the district of Semiriechinsk in Central Asia provided the cradle for the epidemic that broke out in 1338. From there, it probably spread eastwards and southwards to China and India, respectively, and westwards to the Crimea, from thence to the rest of the Old World (Ziegler 1991). The plague might have reached the Nile outflows along its densely populated valley from Egypt through Nubia and Ethiopia, which were also embraced by the second pandemic (McKeown 1988).

The third pandemic started in 1855 in the Chinese province of Yünnan, where troop movements during the war in that area caused a rapid spread of the disease to the southern coast of China (Perry and Fetherston 1997; Simond 1898). Plague reached Hong Kong and Canton in 1894, Bombay in 1898 and, by 1899–1900, steamships had disseminated the disease worldwide (Fig. 9.2). Most scientists believe that plague was introduced relatively recently into America by human beings migrating from Asia. Altogether, there is currently no convincing evidence that the three plague pandemics emerged elsewhere than in Asia.

9.4 Palaeomicrobiology of Plague

Yersinia pestis-specific sequences have now been found, by two independent teams, in human remains dating back to first (Justinian) and second ("Black Death") pandemics (Drancourt et al. 1994; Wiechmann and Grupe 2005; Signoli et al. 1996; Drancourt et al. 1998) (Fig. 9.5). We initially found specific *Y. pestis* sequences, i.e. chromosome-borne *rpo*B and plasmid-borne *pla* genes, in four individuals thought to have died during the 1590 and 1722 plague epidemics in the Marseilles area, but not in seven negative controls. We based our molecular analyses on dental pulp for several reasons, including its susceptibility to septicaemic pathogens, durability, protection against external contamination and ease of manipulation in the laboratory (Drancourt et al. 1998). Using the suicide PCR protocol, in which primers are used only once, we further detected specific sequences in the dental pulp specimens of three individuals thought to have died of plague during the fourteenth century Black

Death pandemic. All negative controls remained negative and original sequences due to point mutations were found in *pla* (Raoult et al. 2000). The typhus agent *Rickettsia prowazekii* and the anthrax agent *Bacillus anthracis* were not detected in these specimens. Furthermore, we were able to detect specific *Y. pestis* sequences in three individuals thought to have died in the fifth–sixth century Justinian plague, along with confirming previous results in Black Death individuals (Drancourt et al. 2004). Later data were independently confirmed by the recovery of two specific *pla* gene sequences from two sixth century individuals from Upper Bavaria, by another research team using total tooth DNA (Wiechmann and Grupe 2005). Interestingly, using the same primers negative results were found when we investigated louse-borne infection in Napoleon's soldiers, and when Papagrigorakis and collaborators investigated the Plague of Athens, which revealed *Salmonella enterica* Typhi (Papagrigorakis et al. 2006). An English team aimed to detect the 16S rRNA gene of *Yersinia pestis* in 61 individuals collected in five burial sites suspected of plague from the thirteenth to the seventeenth century in Northern Europe (Gilbert et al. 2004). Although the authors claimed they failed to detect *Y. pestis*, two specimens yielded sequences that matched enterobacterial 16S rRNA gene sequence, including *Y. pestis*, with 99% and 100% similarity, respectively. Because *Y. pestis*, *Y. pseudotuberculosis* and *Y. enterocolitica* share 100% sequence similarity in the targeted 16S rRNA gene region, the data were non-interpretable but they do not positively exclude the presence of *Y. pestis* in these two particular specimens. Either differences in technical protocols or lack of plague in these individuals may explain discrepancies between studies (Gilbert et al. 2004). A specific 93-bp region of the *Y. pestis caf*1 gene was also detected by PCR and Southern hybridisation in 2/12 seventeenth century individuals suspected of ancient plague and 0/12 controls (Pusch et al. 2004). In these individuals, detection of the *Y. pestis* F1 capsular antigen was achieved in 10/12 individuals but in none of the 12 controls or contemporary soil specimens (Pusch et al. 2004). The fact that molecular evidence of *Y. pestis*-specific DNA sequences have now be obtained by two independent teams, and some original sequences have been identified in agreement with criteria for authenticity in palaeomicrobiology (Drancourt and Raoult 2005), should end the controversy regarding the etiology of historical plagues.

9.5 The Causative Strains of Pandemics

All strains isolated from areas unaffected by plague before the third pandemic are of biotype Orientalis. In ancient enzootic foci, other biotypes were found (Antiqua, Medievalis, Microtus). On the basis of this geographical repartition of biotypes in plague foci, and putative historical data regarding the potential geographical source of past plague epidemics, Devignat hypothesised in 1951 that each of the three known biovars Antiqua, Medievalis, and Orientalis caused the first, second, and third pandemics, respectively (Devignat 1951). This contention was based on the hypothesis that contemporary foci have been the primary sources for each pandemic, a contention without any scientific basis as acknowledged by Devignat himself.

This speculation did not receive any significant confirmation but remained unchallenged and over time became established as a common hypotheses (Achtman et al. 1999; Devignat 1951). Moreover (see above), the original source of plague for the second and the third pandemics was clearly Asian, and the source of the first is disputed between Africa and Asia.

Recent genetic data challenge this hypothesis. Suppression-subtractive hybridisation techniques have demonstrated that Antiqua isolates share a 15,603 bp chromosomal fragment in common with *Y. pseudotuberculosis* and some Microtus isolates (Radnedge et al 2001). Further genome sequencing of two Antiqua isolates refined phylogenetic representation by delineating an Asian Antiqua subgroup 1, closely related to Orientalis, and an African Antiqua subgroup 2, closely related to Medievalis (Chain et al. 2006). Thorough investigation of a 156 isolate *Y. pestis* collection based on synonymous SNPs, VNTR and insertion of IS*100* elements found the *Y. pestis* species to comprise eight populations, which did not match with the one biotype / one pandemic theory (Achtman et al. 2004). Indeed, while one Orientalis population was clearly associated with last pandemic, biovars Antiqua and Medievalis were found to be too polyphyletic to be unambiguously associated with first and second pandemics, respectively (Achtman et al. 2004). Evidence gathered from DNA microarray analysis of genome dynamics in *Y. pestis* indicated that Orientalis evolved in China directly from Antiqua, not from Medievalis (Zhou et al. 2004). Also, micro-evolution analysis of *Y. pestis* clearly indicated a greater diversity of *Y. pestis* isolates in Asia than in Africa, and high diversity is often a good indicator of the geographical source of microbes (Achtman et al. 2004). Altogether, genetic as well as contemporary investigations lead us to consider Asia as the source of the pandemic strain (or strains).

9.6 Identifying the Causative Strains in Human Remains

In order to identify the genotype involved in the three pandemics, we applied MST to dental pulp collected from the remains of eight persons who likely died in the first and second pandemics (Fig. 9.4). In the 46 PCR experiments we performed, we obtained 10 *Y. pestis* sequences in seven of eight persons' remains and no sequences were found in the 51 PCR experiments with negative control teeth of 17 persons ($P < 10^{-4}$). YP1 PCR yielded an amplicon in one of six tested persons; its sequence revealed complete similarity with the homologous region in *Y. pestis* Orientalis over 390 positions. YP8 PCR yielded an amplicon with identical sequence in six of six tested persons, which exhibited 99% sequence similarity with the homologous region in *Y. pestis* Orientalis over 178 positions. YP3 PCR yielded an amplicon in three of seven tested persons; its sequence yielded complete sequence identity with that of the homologous region in *Y. pestis* Orientalis over 364 positions in two persons and a 98% similarity with the homologous region in *Y. pestis* CO92 strain over 283 positions in the last case. This amplicon exhibited

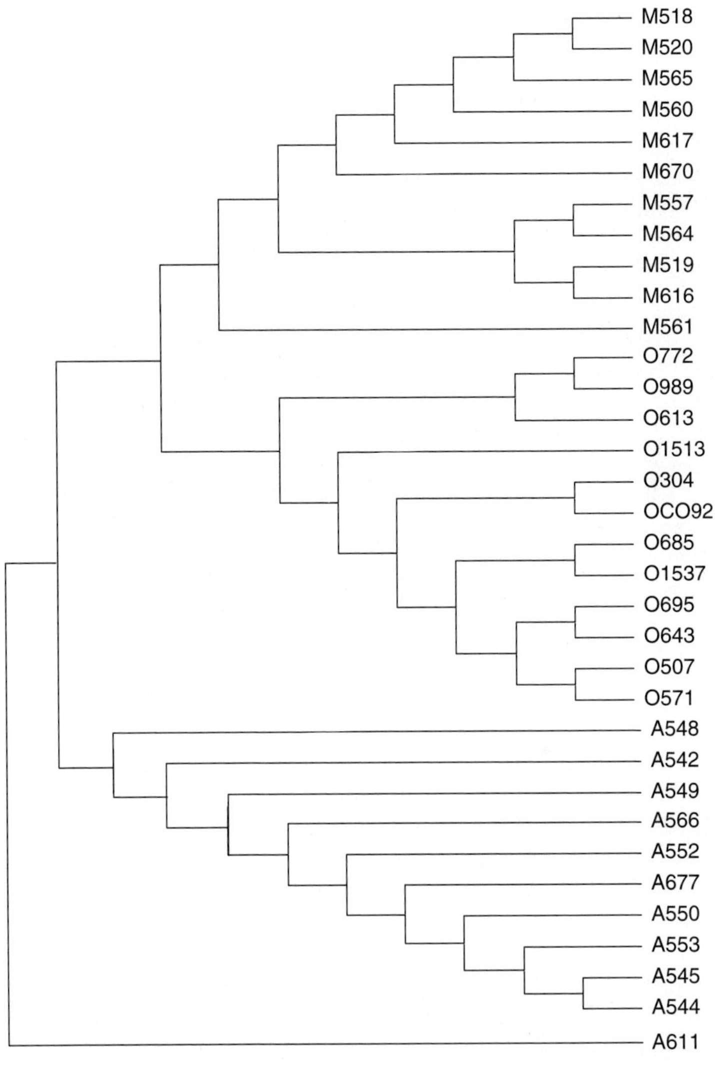

Fig. 9.4 Multispacer sequence typing (MST)-based tree of *Yersinia pestis*. *O* Orientalis biotype, *M* Medievalis, *A* Antiqua, followed by the number of the strain, including reference strain CO92

two specific nucleotide substitutions that were consistently obtained in six clones. MST therefore indicated that the historical strains were more closely related to biotype Orientalis than to the other two biotypes of *Y. pestis* (Drancourt et al. 2004). We further amplified and sequenced the *glp*D gene encoding glycerol-3-phosphate dehydrogenase in five individuals from the two historical pandemics and found the 93-bp deletion reported to be specific for the Orientalis biotype, thus confirming the

Fig. 9.5 A view of the 1720 plague mass grave in Martigues, southern France

MST data (Motin et al. 2002). Palaeomicrobiology data have recently been further augmented by the discovery that a YpF Φ filamentous phage is stably integrated in Orientalis isolates but formed unstable episomes in Antiqua and Medievalis isolates (Derbise et al. 2007). This phage contributes to the pathogenicity of *Y. pestis* in mice and may confer a selective advantage to *Y. pestis* under natural conditions. This illustrates the fact that palaeomicrobiology studies can describe unique characteristics of ancient pathogens, further elucidated by studying modern isolates. At this point, palaeomicrobiology investigations thus indicate that the Orientalis biotype has pandemic potential, whereas other biotypes may have a more limited diffusion potential.

According to the data presented herein, we now hypothesise that the three plague epidemics originated in Asia and were caused by *Y. pestis* biotype Orientalis. We hypothesise that the continuous evolution of the *Y. pestis* genome conferred unique biological properties on biotype Orientalis, including the capacity to promote pandemia, whereas the Antiqua, Medievalis and Microtus biotypes were local variants with limited epidemic potential. Such unique biological properties may be linked to *Y. pestis*–vector relationships or increased capacity to induce septicaemia in its host.

Acknowledgements We acknowledge Michel Signoli for illustration contributions.

References

Advisory Committee (1907) Reports on plague investigations in India. J Hyg 7:323–476

Achtman M, Zurth K, Morelli G, Torrea G, Guiyoule A, Carniel E (1999) *Yersinia pestis* the cause of plague is a recently emerged clone of *Yersinia pseudotuberculosis*. Proc Natl Acad Sci USA 96:14043–14048

Achtman M, Morelli G, Zhu P, Wirth T, Diehl I, Kusecek B, Vogler AJ, Wagner DM, Allender CJ, Easterday WR, Chenal-Francisque V, Worsham P, Thomson NR, Parkhill J, Lindler LE, Carniel E, Keim P (2004) Microevolution and history of the plague bacillus *Yersinia pestis*. Proc Natl Acad Sci USA 101:17837–17842

Anisimov AP, Lindler LE, Pier GB (2004) Intraspecific diversity of *Yersinia pestis*. Clin Microbiol Rev 17:434–464

Chain PS, Carniel E, Larimer FW, Lamerdin J, Stoutland PO, Regala WM, Georgescu AM, Vergez LM, Land ML, Motin VL, Brubaker RR, Fowler J, Hinnebusch J, Marceau M, Medigue C, Simonet M, Chenal-Francisque V, Souza B, Dacheux D, Elliott JM, Derbise A, Hauser LJ, Garcia E (2004) Insights into the evolution of *Yersinia pestis* through whole-genome comparison with *Yersinia pseudotuberculosis*. Proc Natl Acad Sci USA 101:13826–13831

Chain PS, Hu P, Malfatti SA, Radnedge L, Larimer F, Vergez LM, Worsham P, Chu MC, Andersen GL (2006) Complete genome sequence of *Yersinia pestis* strains Antiqua and Nepal516: evidence of gene reduction in an emerging pathogen. J Bacteriol 188:4453–4463

Chanteau S, Ratsifasoamanana L, Rasoamanana B, Rahalison L, Randriambelosoa J, Roux J, Rabeson D (1998) Plague a reemerging disease in Madagascar. Emerg Infect Dis 4:101–104

Deng W, Burland V, Plunkett G III, Boutin A, Mayhew GF, Liss P, Perna NT, Rose DJ, Mau B, Zhou S, Schwartz DC, Fetherston JD, Lindler LE, Brubaker RR, Plano GV, Straley SC, McDonough KA, Nilles ML, Matson JS, Blattner FR, Perry RD (2002) Genome sequence of *Yersinia pestis* KIM. J Bacteriol 184:4601–4611

Derbise A, Chenal-Francisque V, Pouillot F, Fayolle C, Prevost MC, Medigue C, Hinnebusch BJ, Carniel E (2007) A horizontally acquired filamentous phage contributes to the pathogenicity of the plague bacillus. Mol Microbiol 63:1145–1157

Devignat R (1951) Varieties of *Pasteurella pestis*; new hypothesis. Bull World Health Organ 4:247–263

Drancourt M, Raoult D (2005) Palaeomicrobiology: current issues and perspectives. Nat Rev Microbiol 3:23–35

Drancourt M, Aboudharam G, Signoli M, Dutour O, Raoult D (1998) Detection of 400-year-old *Yersinia pestis* DNA in human dental pulp: an approach to the diagnosis of ancient septicemia. Proc Natl Acad Sci USA 95:12637–12640

Drancourt M, Roux V, Dang LV, Tran-Hung L, Castex D, Chenal-Francisque V, Ogata H, Fournier PE, Crubezy E, Raoult D (2004) Genotyping Orientalis-like *Yersinia pestis* and plague pandemics. Emerg Infect Dis 10:1585–1592

Drancourt M, Houhamdi L, Raoult D (2006) *Yersinia pestis* as a telluric human ectoparasite-borne organism Lancet Infect Dis 6:234–241

Enselme J (1969) Commentaries on the great plague of 1348 in Avignon (in French). Rev Lyon Med 17:697–710

Gage KL, Ostfeld RS, Olson JG (1995) Nonviral vector-borne zoonoses associated with mammals in the United States. J Mammalogy 76:695–715

Galimand M, Guiyoule A, Gerbaud G, Rasoamanana B, Chanteau S, Carniel E, Courvalin P (1997) Multidrug resistance in *Yersinia pestis* mediated by a transferable plasmid. New Engl J Med 337:677–680

Gauthier J, Raybaud A (1903) Recherches expérimentales sur le rôle des parasites du rat dans la trasmission de la peste. Rev Hyg 1903:426–438

Gilbert MTP, Cuccui J, White W, Lynnerup N, Titball RW, Cooper A, Prentice MB (2004) Absence of *Yersinia pestis*-specific DNA in human teeth from five European excavations of putative plague victims. Microbiology 150:341–354

Guiyoule A, Grimont F, Iteman I, Grimont PA, Lefevre M, Carniel E (1994) Plague pandemics investigated by ribotyping of *Yersinia pestis* strains. J Clin Microbiol 32:634–641

Houhamdi L, Lepidi H, Drancourt M, Raoult D (2006) Experimental model to evaluate the human body louse as a vector of plague. J Infect Dis 194:1589–1596

Inglesby TV, Dennis DT, Henderson DA, Bartlett JG, Ascher MS, Eitzen E, Fine AD, Friedlander AM, Hauer J, Koerner JF, Layton M, McDade J, Osterholm MT, O'Toole T, Parker G, Perl TM, Russell PK, Schoch-Spana M, Tonat K, Working Group on Civilian Biodefense (2000) Plague as a biological weapon: medical and public health management. JAMA 283:2281–2290

Kitasato S (1894) The bacillus of bubonic plague. Lancet 2:428–430

Le Flèche P, Hauck Y, Onteniente L, Prieur A, Denoeud F, Ramisse V, Sylvestre P, Benson G, Ramisse F, Vergnaud G (2001) A tandem repeats database for bacterial genomes: application to the genotyping of *Yersinia pestis* and *Bacillus anthracis*. BMC Microbiol 1:2

Lowell JL, Wagner DM, Atshabar B, Antolin MF, Vogler AJ, Keim P, Chu MC, Gage KL (2005) Identifying sources of human exposure to plague. J Clin Microbiol 43:650–656

Lucier TS, Brubaker RR (1992) Determination of genome size macrorestriction pattern polymorphism and nonpigmentation-specific deletion in *Yersinia pestis* by pulsed-field gel electrophoresis. J Bacteriol 174:2078–2086

McKeown T (1988) The origins of human diseases. Blackwell, Oxford

Motin VL, Georgescu AM, Elliott JM, Hu P, Worsham PL, Ott LL, Slezak TR, Sokhansanj BA, Regala WM, Brubaker RR, Garcia E (2002) Genetic variability of *Yersinia pestis* isolates as predicted by PCR-based IS100 genotyping and analysis of structural genes encoding glycerol-3-phosphate dehydrogenase (glpD). J Bacteriol 184:1019–1027

Papagrigorakis MJ, Yapijakis C, Synodinos PN, Baziotopoulou-Valavani E (2006) DNA examination of ancient dental pulp incriminates typhoid fever as a probable cause of the Plague of Athens. Int J Infect Dis 10:206–214

Parkhill J, Wren BW, Thomson NR, Titball RW, Holden MT, Prentice MB, Sebaihia M, James KD, Churcher C, Mungall KL, Baker S, Basham D, Bentley SD, Brooks K, Cerdeno-Tarraga AM, Chillingworth T, Cronin A, Davies RM, Davis P, Dougan G, Feltwell T, Hamlin N, Holroyd S, Jagels K, Karlyshev AV, Leather S, Moule S, Oyston PC, Quail M, Rutherford K, Simmonds M, Skelton J, Stevens K, Whitehead S, Barrell BG (2001) Genome sequence of *Yersinia pestis* – the causative agent of plague. Nature 413:523–527

Perry RD, Fetherston JD (1997) *Yersinia pestis* – etiologic agent of plague. Clin Microbiol Rev 10:35–66

Pollitzer R (1954) Plague. World Health Organization, Geneva

Pourcel C, André-Mazeaud F, Neubauer H, Ramisse F, Vergnaud G (2004) Tandem repeats analysis for the high resolution phylogenetic analysis of *Yersinia pestis*. BMC Microbiol 4:22

Procopius A (1914) History of the wars I. Loeb Classical Library, London

Pusch CM, Rahalison L, Blin N, Nicholson GJ, Czarnetzki A (2004) Yersinial F1 antigen and the cause of Black Death. Lancet Infect Dis 4:484–485

Radnedge L, Gamez-Chin S, McCready PM, Worsham PL, Andersen GL (2001) Identification of nucleotide sequences for the specific and rapid detection of *Yersinia pestis*. Appl Environ Microbiol 67:3759–3762

Rakin A, Heesemann J (1995) The established *Yersinia pestis* biovars are characterized by typical patterns of I-CeuI restriction fragment length polymorphism. Mol Gen Mikrobiol Virusol July–Sep:26–29

Raoult D, Aboudharam G, Crubezy E, Larrouy G, Ludes B, Drancourt M (2000) Molecular identification by "suicide PCR" of *Yersinia pestis* as the agent of medieval black death. Proc Natl Acad Sci USA 97:12800–12803

Signoli M, Da Silva J, Léonetti G, Dutour O (1996) Verification of death during the Great Plague of Marseilles: anthropological data from the excavation of the mass grave of l'Observance. C R Acad Sci Paris 322:333–339

Simond PL (1898) La propagation de la peste. Ann Inst Pasteur 10:626–687

Skurnik M, Peippo A, Ervela E (2000) Characterization of the O-antigen gene clusters of *Yersinia pseudotuberculosis* and the cryptic O-antigen gene cluster of *Yersinia pestis* shows that the plague bacillus is most closely related to and has evolved from *Y. pseudotuberculosis* serotype O:1b. Mol Biol Evol 37:316–330

Song Y, Tong Z, Wang J, Wang L, Guo Z, Han Y, Zhang J, Pei D, Zhou D, Qin H, Pang X, Han Y, Zhai J, Li M, Cui B, Qi Z, Jin L, Dai R, Chen F, Li S, Ye C, Du Z, Lin W, Wang J, Yu J, Yang H, Wang J, Huang P, Yang R (2004) Complete genome sequence of *Yersinia pestis* strain 91001: an isolate avirulent to humans. DNA Res 11:179–197

Wiechmann I, Grupe G (2005) Detection of *Yersinia pestis* DNA in two early medieval skeletal finds from Aschheim (Upper Bavaria 6th century A.D.). Am J Phys Anthropol 126:48–55

Wren BW (2000) Microbial genome analysis: insights into virulence host adaptation and evolution. Nat Rev Genet 1:30–39

Yersin A (1894) La peste bubonique à Hong-Kong. Ann Inst Pasteur 8:662–667

Zhou D, Tong Z, Song Y, Han Y, Pei D, Pang X, Zhai J, Li M, Cui B, Qi Z, Jin L, Dai R, Du Z, Wang J, Guo Z, Wang J, Huang P, Yang R (2004) Genetics of metabolic variations between *Yersinia pestis* biovars and the proposal of a new biovar microtus. J Bacteriol 186:5147–5152

Ziegler P (1991) The Black Death. Sutton, Wolfeboro Falls, NH

Chapter 10
Typhoid Fever Epidemic in Ancient Athens

Manolis J. Papagrigorakis(✉), Christos Yapijakis, and Philippos N. Synodinos

Abstract Molecular evidence, resulting from investigation and analysis of ancient DNA, has identified the presence of *Salmonella enterica* serovar Typhi in victims of the Plague of Athens, thereby incriminating typhoid fever as a likely cause of the epidemic. Current clinical and epidemiological scientific data, related to modern-day typhoid, correlate well to the signs and symptoms of the disease as Thucydides has described them, whereas their apparent differences may be reasonably explained. The most striking hypothesis is that the ancient *S. typhi* strain may constitute the ancestral original strain of the pathogen, capable of affecting both human and animal hosts. The genomic evolution of the ancient *Salmonella typhi* strain over time may provide a satisfactory explanation for the diminished morbidity and the varying clinical symptomatology of modern-day typhoid fever. Further investigations, implementing DNA sequencing techniques of the ancient strain of *S. enterica*, may elucidate its genetically determined differences from its modern counterpart, thus facilitating new approaches to preventing or treating typhoid fever epidemics.

10.1 The Plague of Athens – Historical Background

At many historical crossroads, major epidemics have been shown to have influenced the rise and fall of several great civilisations. Such an epidemic is, undoubtedly, the one that literally decimated the population of the city-state of Athens around 430–426 B.C. – generally known as the Plague of Athens [Thucydides §2.47–2.54 (note that all references to Thucydides' description are taken from C.F. Smith's 1919 translation (Thucydides 1919)]. This epidemic was a decisive factor that changed the balance of power, thereby determining the outcome of the Peloponnesian War of Athens against Sparta, thus ending the Golden Age of Pericles (Thucydides §2.47–2.54; Longrigg 1980; Soupios 2004). As a direct

Manolis J. Papagrigorakis
Department of Orthodontics, Dental School, University of Athens, 2 Thivon Str., 11527 Goudi, Athens, Greece, and Museum of Craniofacial Evolution and Dental History, Dental School, University of Athens, 2 Thivon Str., 11527 Goudi, Athens, Greece, E-mail: manjpap@dent.uoa.gr

D. Raoult and M. Drancourt (eds.), *Paleomicrobiology: Past Human Infections.*

consequence, Athenian predominance in the Mediterranean Sea declined, thereby greatly affecting the route of history.

The Plague broke out during the siege of Athens by the Spartan army early in the summer of 430 B.C. (Thucydides §2.47–2.54). After a brief remission in 428 B.C., the epidemic reemerged in the winter of the following year and lasted until the winter of 426 B.C. About one-third of the Athenian population, one-fourth of their army and their charismatic leader Pericles, perished in the epidemic.

Until recently, all data pertaining to the Plague and its clinical characteristics were based on the account of the epidemic as reported by the fifth century B.C. Athenian historian Thucydides, who himself was taken ill with the Plague but recovered. In his famous history of the Peloponnesian War, Thucydides gave a detailed description of the epidemic (Thucydides §2.47–2.54) that has since formed the basis of several hypotheses regarding its cause. Thucydides' records describe major signs and symptoms of the disease as well as other associated events and behaviour of affected and non-affected Athenians. The validity and reliability of these narrations is taken for granted.

10.2 Hypotheses on the Cause of the Plague of Athens

Although extensively informative, the narration of Thucydides regarding the Plague of Athens does not correspond exactly to the characteristics of any single disease as it is known today. This fact has prompted various authors to suggest about 30 possible pathogens that might putatively be implicated in the emergence and spread of the disease, based on Thucydides' description of its signs and symptoms. These included the causative agents of smallpox, Lassa fever, measles, scarlet fever, tuberculosis (*Mycobacterium tuberculosis*), epidemic typhus (*Rickettsia prowazekii*), anthrax (*Bacillus anthracis*), typhoid fever (*Salmonella enterica* serovar Typhi), plague (*Yersinia pestis*), Ebola, and even staphylococcal toxic shock syndrome as a complication of influenza (Shrewsbury 1950; Page 1953; Littman and Littman 1969; Langmuir et al. 1985; Holladay 1986; Scarrow 1988; McSherry and Kilpatrick 1992; Olsen et al. 1996, 1998; Holden 1996; Perry and Fetherston 1997; Durack et al. 2000; Cunha 2004). Nevertheless, each one of the proposed "most likely" causative agents of the Plague of Athens explained only a percentage of the epidemic characteristics described by Thucydides.

In any case, the mystery of the Plague that precipitated the end of the Golden Age of Athens has, until recently, remained unresolved due to the lack of definitive archeological and biological evidence.

10.3 A Mass Burial Site of Plague Victims

The skeletal material necessary for an objective investigation of the cause of the Plague of Athens was provided by a recent excavation (1994–1995) conducted in Kerameikos' ancient cemetery of Athens.

In this excavation, a mass burial site was discovered in the outskirts of Kerameikos (Baziotopoulou-Valavani 2002). The mass grave was a simple 6.5 m-long pit of rather irregular shape that contained the remains of at least 150 individuals of various ages (men, women and children). The dead bodies were laid in a disorderly manner, most of them in outstretched positions, but several were placed with their feet directed towards the pit centre and their heads towards the circumference (Baziotopoulou-Valavani 2002). They formed more than five successive layers, without any intervening soil between them. At the lower level, the deceased were more distant from each other, although their manner of placement was as disordered as in the upper layers. It seemed that more care during burial had been taken at the lower levels of the mass grave, while at the upper levels the dead were virtually heaped one upon the other (Baziotopoulou-Valavani 2002). In the upper layer, eight pot burials of infants were found, indicating that special care was taken for them in contrast to the careless piling of adults in the same pit (Baziotopoulou-Valavani 2002).

A few offerings of about 30 small vases were found scattered among the bodies of the lower layers of the grave (Baziotopoulou-Valavani 2002). The quality and quantity of these offerings were extremely poor and absolutely disproportionate to such a large number of buried people. Most of the discovered vases were dated at around 430 B.C., using accepted chronological techniques (Baziotopoulou-Valavani 2002). The chronology of the few burial offerings as well as the hasty and impious manner of burying about 150 dead were associated with the outbreak of the Plague of Athens during the first years of the Peloponnesian War, between 430–426 B.C. (Baziotopoulou-Valavani 2002).

The absence of archaeological evidence relating to the victims of the Plague is due to the fact that, in most cases, the relatives of the deceased usually undertook other ways of burials, such as cremations or individual inhumations (Thucydides §2.47–2.54).

The mass burial of Kerameikos evidently did not have a monumental character. It had been completed in such a hasty, improper and uncommonly impious manner, that any possibility of addressing the dead as victims of war was therefore excluded (Baziotopoulou-Valavani 2002). Instead, the most likely explanation is that the authorities of the City of Athens hastily buried a large number of poor and hapless dead Plague victims as a means of protecting its still surviving population from the epidemic (Baziotopoulou-Valavani 2002). Mass graves are rather rare in the ancient Greek world, and the few known such examples in the Classical period have been connected to extreme circumstances such as the outbreak of lethal, epidemic plague-like diseases.

In this case, a large number of dead bodies were thrown one upon the other (rather than buried) in ways that were dictated primarily by the shape and size of the irregular and roughly dug pit.

Therefore, the mass burial pit of Kerameikos offered the opportunity for a biomedical evidence-based approach towards resolving the mystery of the agent that caused the Plague of Athens, through the study of the recovered human skeletal remains.

10.4 DNA Extraction From Teeth of Plague Victims

The molecular detection of microbial DNA sequences in ancient skeletal material has made possible the retrospective diagnosis of ancient infectious diseases (Pääbo 1989; Taylor et al. 1996, 1999; Taubenberger et al. 1997; Nerlich et al. 1997; Kolman et al. 2000; Raoult et al. 2000, 2006). Recovered DNA fragments of ancient microorganisms may be enzymatically amplified by various methods of polymerase chain reaction (PCR) and consequently sequenced to assess their similarity with their modern-day counterparts deposited in electronic databases (Taylor et al. 1996, 1999; Nerlich et al. 1997; Taubenberger et al. 1997; Kolman et al. 2000; Raoult et al. 2000, 2006; Cunha 2004). In addition to highly accurate molecular methods, extreme preventive measures must be applied in order to minimise the risk of false-positive results due to sample contamination by previously attempted analyses or naturally occurring microorganisms.

In the case of the Plague of Athens, the material of choice was intact teeth (as observed macroscopically and verified by X-radiographs) randomly collected from three different Plague victims of the Kerameikos mass grave, because DNA remnants from systemic pathogens causing bacteraemia had previously been shown to be present in ancient dental pulp (Holden 1998; Raoult et al. 2000, 2006). By virtue of its good vascularisation, durability and natural sterility, dental pulp is considered well protected from any external contamination. Only in case of bacteraemia, would the ancient dental pulp have trapped genetic material of the pathogenic microorganism in amounts adequate for study (Holden 1998; Aboudharam et al. 2000; Raoult et al. 2000, 2006; Drancourt and Raoult 2002).

Since no other dental archaeological material was available, matching the historical time and location attributes of the material under study, two modern intact teeth served as negative controls against any false-positive amplification of distantly related human genomic sequences (Papagrigorakis et al. 2006a). In addition, a soil sample washed from ancient teeth served as a negative control of external contamination during DNA extraction (Papagrigorakis et al. 2006a). No positive controls were included, thereby excluding any possible contamination of the ancient material by DNA from the microbes under study. For the same reason, the steps of DNA extraction, PCR amplification and DNA sequencing reactions were performed in three laboratories located in different buildings (Papagrigorakis et al. 2006a). None of the pathogens under study or their respective primer sequences had ever been introduced into any of these laboratories, thus minimising the risk of false-positive results due to contamination of the ancient material. Furthermore, in order to avoid any bias of the examiner, no data regarding the origin of the teeth or the actual purpose of the analysis were available to the staff of the laboratories that participated in the study (Papagrigorakis et al. 2006a).

Ancient and modern teeth were thoroughly washed and fractured longitudinally. The remnants of the dental pulp, which were powdery in ancient teeth, were scraped off and transferred into sterile tubes. Using a Forensic DNA Trace kit (Nucleospin DNA Trace, Macherey-Nagel, Düren, Germany), total dental pulp

DNA was isolated from all teeth under study as well as from a soil sample washed from ancient teeth (Papagrigorakis et al. 2006a).

10.5 "Suicide" PCR Attempts for Candidate Pathogens

In order to assess the preservation of DNA in ancient dental pulp, PCR amplification of a human genomic sequence in extracted DNA samples was attempted. A region of the coagulation factor V gene was amplified in all samples extracted from ancient and modern teeth, but not in the sample extracted from the soil wash (Papagrigorakis et al. 2006a).

The investigation of the Plague-causing pathogen was effected by two consecutive rounds of "suicide" PCR amplification performed simultaneously in all DNA samples corresponding to all studied teeth and the soil wash (Papagrigorakis et al. 2006a). The previously described "suicide PCR" method (Raoult et al. 2000) permitted only a single use of each primer pair, targetted successively at candidate microbial DNA sequences, until a product of the expected size was obtained and visualised by ethidium bromide staining after agarose gel electrophoretic analysis. The identity of this product then had to be confirmed by DNA sequencing.

The presence of seven putative causative agents of the Plague of Athens was randomly and successively investigated. No PCR product was yielded in "suicide" reactions of ancient DNA samples and controls using primers for sequences of the agents of plague (*Yersinia pestis*), typhus (*Rickettsia prowazekii*), anthrax (*Bacillus anthracis*), tuberculosis (*Mycobacterium tuberculosis*), cowpox (*Cowpox virus*) and cat-scratch disease (*Bartonella hensellae*) (Papagrigorakis et al. 2006a).

The seventh such attempt, targetted at a sequence of the agent of typhoid fever (*Salmonella enterica* serovar Typhi), eventually yielded, in all three ancient teeth, a product of the expected size, corresponding to a 322 bp fragment containing parts of the *osmC* (encoding osmotically inducible protein C) and *clyA* (encoding cytolysin A) genes (Papagrigorakis et al. 2006a). DNA sequencing confirmed that the sequence of the PCR product was highly homologous (96%) to parts of both genes of *S. enterica* Typhi, while the intervening sequence displayed lower sequence homology (80%) (Papagrigorakis et al. 2006a). On the contrary, no product was obtained using the same primers under the same laboratory conditions on the negative controls (modern teeth and soil wash) (Papagrigorakis et al. 2006a). In order to confirm the putative presence of *S. enterica* Typhi genomic sequences in the examined ancient dental pulp samples, another "suicide" PCR was attempted using a pair of primers targeted at the *narG* gene (encoding the alpha chain of nitrate reductase 1). (Parkhill et al. 2001; Deng et al. 2003). The expected PCR product of 360 bp was obtained in all three ancient DNA samples but in none of the three negative controls. DNA sequencing revealed that it shared 93% sequence homology with the respective *narG* gene of *S. enterica* Typhi (Parkhill et al. 2001; Deng et al. 2003; Papagrigorakis et al. 2006a). All positive PCR amplifications of ancient DNA were repeated independently three times in two different laboratories by two different specialists (Papagrigorakis et al. 2006a).

10.6 An Ancestral Strain of *Salmonella enterica* Serovar Typhi?

The identified microbial sequences that existed in the ancient dental pulp of Plague victims were clearly highly homologous to modern sequences of the typhoid fever agent, but they did not match exactly with them. Was this an artifact either due to chemical decomposition of some nucleotides over time or due to amplification of patchy DNA fragments of various origins? Alternatively, did this result suggest that the DNA sequences identified belonged to an ancient strain of *S. enterica* serovar Typhi?

A 240 bp region of the *narG* sequence, which was clearly detected with no background in both strands of all samples after direct sequencing and cloned PCR-sequencing, was further analysed. It contained 28 nucleotide alterations (involving most possible nucleotide swaps) from the present day Ty2 strain of *S. enterica* Typhi (Papagrigorakis et al. 2006a). The great majority of these alterations (25/28) were single-base polymorphisms in the third position of the codon that did not alter its genetic meaning, practically excluding the possibilities of either accidental chemical damage of particular nucleotides or amplification of a chimaeric PCR product (Papagrigorakis et al. 2006a). Only three nucleotide alterations were single-base missense mutations resulting in amino acid changes (Met85Leu, Met118Ile and Leu120Met); the effect of these mutations on the spatial conformation and activity of the *narG* gene product, which is involved in anaerobic respiration, is unclear (Papagrigorakis et al. 2006a).

These data ascertain that an ancient strain of the typhoid fever agent *S. enterica* serovar Typhi was present in the dental pulp of three randomly selected individuals buried in a mass grave of about 150 individuals, which was dated to the era of the Plague of Athens. Both the fact that *S. enterica* Typhi sequences of three genes were independently amplified in three different Plague victims, and the fact that six alternative candidate agents previously investigated as candidate causes of the Plague were not identified, further reinforce the above assumption. Any possibility that the detected PCR products could have resulted from a modern and currently unknown free-living soil bacterium instead of an ancient one was excluded, since, under the same conditions, application of the same primers to soil washed from the ancient teeth failed to yield any product (Papagrigorakis et al. 2006a, 2006b). The fact that the ancient microbial DNA sequences were preserved for more than 24 centuries in good enough condition for molecular detection and analysis in all three studied teeth might possibly reflect the presence of a large amount of *S. enterica* Typhi cells due to bacteraemia. The typhoid fever agent is indeed a deadly septicaemic pathogen, and dental pulp is known to be appropriate for the detection of bacteraemic pathogens (Glick et al. 1991; Raoult et al. 2000, 2006; Aboudharam et al. 2000; Parry et al. 2000).

In conclusion, it seems that a strain of *S. enterica* serovar Typhi (or a bacterial species very closely related to it, if not *S. enterica* Typhi *stricto sensu*) was clearly involved in the epidemic that devastated Athens in 430–426 B.C. (Papagrigorakis et al. 2006a). The 93–96% homology of the ancient DNA sequences to the corresponding sequences of a present day strain of the typhoid fever agent is regarded as high enough to allow the above conclusion to be considered as safe. If another, presently unknown pathogen (and not an ancestral strain of *S. enterica*) was the actual cause of the Plague of Athens, it would have to be closely related to *S. enterica*, and definitely closer than *S. typhimurium* or *E. coli.*

Salmonella typhimurium, which is the closest known relative of *S. enterica* Typhi (McClelland et al. 2001), shared less than 91% homology with the ancient DNA sequences in two genes (*osmC* and *narG)* and even lacks the entire *clyA* gene (Papagrigorakis et al. 2006a; Oscarsson et al. 2002). Other bacteria showing 80–88% homology to the ancient DNA sequences were *Escherichia coli*, *Erwinia carotovora* and *Shigella flexneri* (Papagrigorakis et al. 2006a).

The identified genomic differences (most of which do not alter the codon meaning) between the recovered DNA from teeth of Kerameikos and present day strains of *S. enterica* provide further clear evidence that the recovered microbial DNA probably belongs to an ancestral strain of *S. enterica* (Papagrigorakis et al. 2006a, 2007).

10.7 Thucydides' Narration Revisited

In order to consider an infectious disease as a likely cause of the Plague of Athens, first of all it must have existed at that time (Cunha 2004). Certainly, ancient descriptions of infectious diarrhoeas and dysentery imply that typhoid fever was an endemic problem in the ancient world (Lim and Wallace 2004). Interestingly, Hippokrates, who also lived in the fifth century, described very accurately the symptomatology of typhoid fever, although he used the name "typhus" instead.

Even though some parts of Thucydides' history were written retrospectively (up to a decade after the recorded facts), it is generally assumed that his narrations are reliable and valid. Taking into account the fact that the historian was not a physician, it is assumed that no key clinical features were omitted, thus the description of the epidemic was as accurate as possible at the time (Durack et al. 2000). Also Thucydides was a keen observer and a careful recorder of events, as well as himself being a victim of the disease, he may not have been able to weigh the relative significance of the variable clinical manifestations of the Plague. Although the historian may have stressed trivial signs and symptoms at the expense of important ones (Durack et al. 2000), the molecular diagnosis of typhoid fever is consistent with some of the key characteristics and clinical features reported by Thucydides (Thucydides §2.47–2.54; Page 1953; Cunha 2004):

1. The sudden outbreak of the disease in the city of Athens has been linked either to incoming travellers (carriers of the disease) from other countries like Egypt to the port of Piraeus, or to the speculated poisoning of the water reservoirs by the Spartans (Thucydides §2.47–2.54). Both etiologies apply to what is observed

even today regarding ways of contracting and spreading typhoid fever in visitors and residents in areas of developing countries endemic for typhoid fever (Hoffner et al. 2000; Parry et al. 2002; Connor and Schwartz 2005; Bhan et al. 2005).

2. The crowded and unsanitary conditions in besieged Athens of 430 B.C. (Thucydides §2.47–2.54) undoubtedly must have favoured the spread of the epidemic, as is usually the case in modern-day typhoid epidemics in developing countries (Parry et al. 2002; Connor and Schwartz 2005; Bhan et al. 2005), especially where the water supplies are contaminated by the pathogen or poor housing does not facilitate personal hygiene (Black et al. 1985; Luby et al. 1998; Mermin et al. 1999; Gasem et al. 2001). These conditions were progressively made even worse for the besieged ancient Athenians by the continuing influx of refugees from the countryside into the city, where they lived in crowded conditions and in close contact with the sick (Thucydides §2.47–2.54).
3. The contagious nature of the disease is emphasised by the fact that relatives, friends and doctors contracted the disease by coming into close contact with affected people (Thucydides §2.47–2.54). Social or physical contact with patients with typhoid fever has also been identified as a risk factor for contracting the disease today (Luxemburger et al. 2001).
4. Thucydides also mentions that all other cases of sickness in the ancient Athens of 430 B.C. ended in the Plague (Thucydides §2.47–2.54). In modern times also, contracting typhoid is more likely in patients with weaker constitutions, e.g. immunosuppressed patients (Hoffner et al. 2000; Bhan et al. 2002).
5. No remedy was found even for the best attended Plague-affected Athenians (Thucydides §2.47–2.54). This is not surprising since the only known effective treatment of typhoid fever is the administration of specific antibiotic drugs (Hoffner et al. 2000; Parry et al. 2002; Connor and Schwartz 2005; Bhan et al. 2005), which of course were not available in ancient Athens.
6. Prevention of contracting the disease was not possible; in modern times this is achieved by vaccinating the entire population of typhoid endemic areas (Parry et al. 2002; Connor and Schwartz 2005; Bhan et al. 2005).
7. The major clinical features of the Plague of Athens, as reported by Thucydides, indicate an epidemic of acute onset and rapid progression that ended fatally for most of the affected persons, decimating about one-third of Athenians, including their charismatic leader, Pericles (Thucydides §2.47–2.54). This is also in accord with a diagnosis of typhoid fever, since even though today's worldwide mortality rate from the disease is 1%, it may rise to up to 30–50% in endemic areas, especially if antibiotic treatment is delayed (Hook 1984; Punjabi et al. 1988; Rogerson et al. 1991; Parry et al. 2000).
8. The molecular diagnosis of typhoid fever is consistent with most of the key clinical features reported by Thucydides (Thucydides §2.47–2.54), including the prolonged and violent attacks of headache and fever, the cough and the sore throat, as well as the subsequent abdominal pain, diarrhoea and rash (Hoffner et al. 2000; Cunha 2004). Some other features of Thucydides' report, including confusion, apathetic behaviour (Thucydides §2.47–2.54) may be observed, though much more rarely, in some modern cases (Hoffner et al. 2000).

Other features of the disease, as cited in Thucydides' work (Thucydides §2.47–2.54), are inconsistent with the typical present-day form of typhoid fever (House et al. 2001; Parry et al 2002; Bhan et al. 2005). These include its reported acute onset and its clinical symptoms of conjuctivitis, development of profound ulcers, gangrene of the extremities, and a sensation of intense internal heat (Thucydides §2.47–2.54). These inconsistencies may be attributed to a possible evolution of the typhoid fever pathogen over time, which means that the disease may not manifest itself in the same fashion today, and that it was present in a much more aggressive form in the past (Soupios 2004; Cunha 2004). Alternatively, the concurrent presence of a plurality of infectious diseases in besieged Athens of 430–426 B.C. cannot yet be excluded, allowing for the variable clinical manifestations of Thucydides' report of the Plague (Durack et al. 2000; Cunha 2004). It would have next to impossible for Thucydides or any other observer to distinguish between two or more such diseases at that time (Durack et al. 2000).

The fact that the residents of Athens of 430 B.C. showed no resistance to the advent and the rapid spread of the epidemic (Thucydides §2.47–2.54) indicates that there was no immunisation of the population by any former introduction of the disease, while at the same time the Athenians were probably malnourished due to the siege of their city by the Spartans (Soupios 2004; Cunha 2004). This may also be an explanation of why the severe complications of typhoid fever, such as cardiovascular, pulmonary, respiratory, central nervous system, neuropsychiatric, hepatobiliary, genitourinary, haematologic and other symptoms, that are sometimes observed in neglected or immuno-suppressed cases of modern times (Parry et al. 2002; Huan and DuPont 2005), were typical of the characteristic clinical features of the Plague of Athens (Thucydides §2.47), where no effective treatment was available.

Death from the Athens' epidemic ensued on the 7th–9th day from initiation of symptoms (Thucydides §2.47–2.54), whereas in modern times, in untreated cases the disease ends fatally due to the development of severe complications after 2–3 weeks (Cunha 2004). In addition, in modern typhoid outbreaks, the case fatality rate is higher among the very young and elderly patients (Stuart and Pullen 1946; Parry et al. 2002), whereas in ancient Athens strong and weak constitutions alike were affected (Thucydides §2.47–2.54). This may be attributed to the general immunisation to typhoid epidemics of the population over time, as people were repeatedly subjected to successive attacks of the disease (Shrewsbury 1950). The same conclusion is also reached through the analysis of Thucydides' account of the disease, as those ancient Athenians that recovered were never attacked twice with the same morbidity and mortality (Thucydides §2.47–2.54). Besides, in modern times, although typhoid fever relapses in 5–10% of cases, the second wave usually occurs in milder clinical form, thus verifying the immunising effect of the first attack (Parry et al. 2002; Bhan et al. 2005). Nevertheless, even today typhoid fever is a major health problem on a global scale. Every year there are about 20 million new cases that lead to about 600,000 deaths in the developing world, where overpopulation, inadequate water supplies and hygiene, as well as poor access to health services allow epidemics to spread with tragic results.

Last, but not least, one of the most outstanding features of Thucydides' account of the Plague of Athens' is the reported contraction of the disease by animals (Thucydides §2.47–2.54). This is very interesting since today typhoid fever is strictly adapted to humans (Parry et al. 2002; Bhan et al. 2005). Not excluding the possibility of a combination of diseases occurring simultaneously in ancient Athens (Durack et al. 2000), this finding may fit a working hypothesis of the intriguing possibility that this specific strain of *S. enterica* Typhi, incriminated as the cause of the Plague of Athens, was an ancient strain that was not adapted to human hosts only (Papagrigorakis et al. 2006b), whose existence has been anticipated (Kidgell et al. 2002).

10.8 Future Prospects

Typhoid fever almost certainly played a part in causing the Plague of Athens, either exclusively or in combination with another (and so far unknown) infection. Genomic differences between ancient and present day *S. enterica* serovar Typhi strains, such as those already identified in the DNA samples of Kerameikos, may offer in the future some reasonable explanation for the differences in clinical features observed between the Plague of Athens and the present day form of typhoid fever (Black et al. 1985).

Nine criteria for validating ancient DNA studies have been proposed (Cooper and Poinar 2000): (1) a physically isolated work area, (2) control amplifications, (3) appropriate molecular behaviour of PCR products, (4) reproducibility of results, (5) cloning of amplified products, (6) survival of associated human DNA remains, (7) independent replication by sequencing in independent laboratories, (8) biochemical preservation studies of DNA, and (9) quantitation of copy number of target DNA using competitive PCR. So far, the first seven criteria have been met for the Plague of Athens, while the remaining two will follow in subsequent studies.

Future prospects for research include the investigation of more Plague victims from Kerameikos, both for *S. enterica* Typhi sequences and those of other candidate pathogens. In addition, a major challenge is the precise genetic characterisation of the ancient strain of the typhoid fever agent, which might lead not only to understanding of its aggressiveness 24 centuries ago, but also to possibly develop animal models of typhoid fever, an important tool to research and combat this disease.

Studying the historical aspects of infectious diseases can be extremely useful in several disciplines. This study sheds light on one of the most debated enigmas in medical history and therefore may contribute to many scientific fields. Archaeology, history, microbiology, palaeopathology, certain fields of medicine, anthropology and even genetics, molecular biology and biology of evolution are clearly implicated in such matters and can benefit from relevant studies.

References

Aboudharam G, Lascola B, Raoult D, Drancourt M (2000) Detection of *Coxiella burnetti* DNA in dental pulp during experimental bacteremia. Microb Pathog 28:249–254

Baziotopoulou-Valavani EA (2002) Mass burial from the Cemetery of Kerameikos. In: Stamatopoulou M, Yeroulanou M (eds) Excavating classical culture. Recent archaeological discoveries in Greece. Studies in Classical Archaeology I. BAR International Series 1031. Archaeopress, Oxford, pp 187–201

Bhan MK, Bahl R, Sazawal S, Sinha A, Kumar R, Mahalanabis D, Clemens JD (2002) Association between *Helicobacter pylori* infection and increased risk of typhoid fever. J Infect Dis 186:1857–1860

Bhan MK, Bahl R, Bhatnagar S (2005) Typhoid and paratyphoid fever. Lancet 366:749–762

Black RE, Cisneros L, Levine MM, Banfi A, Lobos H, Rodriguez H (1985) Case-control study to identify risk factors for paediatric endemic typhoid fever in Santiago, Chile. Bull World Health Organ 63:899–904

Connor BA, Schwartz E (2005) Typhoid and paratyphoid in travelers. Lancet Infect Dis 5:623–628

Cooper A, Poinar HN (2000) Ancient DNA: do it right or not at all. Science 289:1139

Cunha BA (2004) The cause of the plague of Athens: plague, typhoid, typhus, smallpox or measles? Infect Dis Clin N Am 18:29–43

Deng W, Liou SR, Plunkett G III, Mayhew GF, Rose DJ, Burland V, Kodoyianni V, Schwartz DC, Blattner FR (2003) Comparative genomics of *Salmonella enterica* serovar Typhi Strains Ty2 and CT18. J Bacteriol 185:2330–2337

Drancourt M, Raoult D (2002) Molecular insights into the history of plague. Microbes Infect 4:105–109

Durack DT, Littman RJ, Benitez RM, Mackowiak PA (2000) Hellenic holocaust: a historical clinico-pathologic conference. Am J Med 109:391–397

Gasem MH, Dolmans WM, Keuter MM, Djokomoeljanto RR (2001) Poor food hygiene and housing as risk factors for typhoid fever in Semarang, Indonesia. Trop Med Int Health 6:484–490

Glick M, Trope M, Bagasra O, Pliskin ME (1991) Human immunodeficiency virus infection of fibroblasts of dental pulp in seropositive patients. Oral Surg Oral Med Oral Pathol 71:733–736

Hoffner RJ, Slaven E, Perez J, Magana RN, Henderson SO (2000) Emergency department presentations of typhoid fever. J Emerg Med 4:317–321

Holden C (1996) Athenian plague probe. Science 274:1307

Holden C (1998) Old dental pulp points to plague. Science 282:619

Holladay AJ (1986) The Thucydides syndrome: another view. N Engl J Med 15:1170–1173

Hook EW (1984) Typhoid fever today. N Engl J Med 310:116–118

House D, Bishop A, Parry CM, Dougan G, Wain J (2001) Typhoid fever: pathogenesis and disease. Curr Opin Infect Dis 14:573–578

Huan DB, DuPont HI (2005) Problem pathogens: extra-intestinal complications of *Salmonella enterica* serotype typhi infection. Lancet Infect Dis 5:341–348

Kidgell C, Reichard U, Wain J, Linz B, Torpdahl M, Dougan G, Achman M (2002) *Salmonella typhi*, the causative agent of typhoid fever, is approximately 50,000 years old. Infect Genet Evol 2:39–45

Kolman CJ, Centurion-Lara A, Lukehart SA, Owsley DW, Tuross N (2000) Identification of *Treponema pallidum* subspecies pallidum in a 200-year-old skeletal specimen. Infect Dis 180:2060–2063

Langmuir AD, Worthen TD, Solomon J, Ray CG, Petersen E (1985) The Thucydides syndrome: a new hypothesis for the cause of the plague of Athens. N Engl J Med 313:1027–1030

Lim ML, Wallace MR (2004) Infectious diarrhea in history. Infect Dis Clin N Am 18:261–274

Littman RJ, Littman ML (1969) The Athenian plague: smallpox. Trans Am Philol Assoc 100:261–275

Longrigg J (1980) The great plague of Athens. Hist Sci 18:209–225

Luby SP, Faizan MK, Fisher-Hoch SP, Syed A, Mintz ED, Bhutta ZA, McCormick JB (1998) Risk factors for typhoid fever in an endemic setting, Karachi, Pakistan. Epidemiol Infect 120:129–138

Luxemburger C, Chau MC, Mai NL, Wain J, Tran TH, Simpson JA, Le HK, Nguyen TTT, White NJ, Farrar JJ (2001) Risk factors for typhoid fever in the Mekong Delta, southern Viet Nam: a case-control study. Trans R Soc Trop Med Hyg 95:19–23

McClelland M, Sanderson KE, Spieth J, Clifton SW, Latreille P, Courtney L, Porwollik S, Ali J, Dante M, Du F, Hou S, Layman D, Leonard S, Nguyen C, Scott K, Holmes A, Grewal N, Mulvaney E, Ryan E, Sun H, Florea L, Miller W, Stoneking T, Nhan M, Waterston R, Wilson RK (2001) Complete genome sequence of *Salmonella enterica* serovar Typhimurium LT2. Nature 413:852–856

McSherry J, Kilpatrick R (1992) The Plague of Athens. J R Soc Med 85:713

Mermin JH, Villar R, Carpenter J, Roberts L, Samaridden A, Gasanova L, Lomakina S, Bopp C, Hutwagner L, Mead P, Ross B, Mintz ED (1999) A massive epidemic of multidrug-resistant typhoid fever in Tajikistan associated with consumption of municipal water. J Infect Dis 179:1416–1422

Nerlich A, Haas CJ, Zink A, Szeimies U, Hagedom HG (1997) Molecular evidence for tuberculosis in an ancient Egyptian mummy. Lancet 350:1404

Olson PE, Hames CS, Benenson AS, Genovese EN (1996) The Thucydides syndrome: *Ebola* deja vu? (or Ebola reemergent?). Emerg Infect Dis 2:155–156

Olson PE, Benerson AS, Genovese EN (1998) *Ebola*/Athens revisited. Emerg Infect Dis 4:134

Oscarsson J, Westermark M, Löfdahl S, Olsen B, Palmgren H, Mizunoe Y, Wai SN, Uhlin BE (2002) Characterization of a pore-forming cytotoxin expressed by *Salmonella enterica* serovars Typhi and Paratyphi A. Infect Immun 10:5759–5769

Pääbo S (1989) Ancient DNA: extraction, characterization, molecular cloning and enzymatic amplification. Proc Natl Acad Sci USA 86:1939–1943

Page DL (1953) Thucydides' description of the great plague. Classical Q 3:97–119

Papagrigorakis MI, Yapijakis C, Synodinos PN, Baziotolpoulou-Valavani E (2006a) DNA examination of ancient dental pulp incriminates typhoid fever as a probable cause of the Plague of Athens. Int J Infect Dis 10:206–214

Papagrigorakis MI, Yapijakis C, Synodinos PN, Baziotolpoulou-Valavani E (2006b) Insufficient phylogenetic analysis may not exclude candidacy of typhoid fever as a probable cause of the Plague of Athens (reply to Shapiro et al.). Int J Infect Dis 10:335–336

Papagrigorakis MJ, Synodinos PN, Yapijakis C (2007) Ancient typhoid epidemic reveals possible ancestral strain of *Salmonella enterica* serovar Typhi. Infect Genet Evol 7:126–127

Parkhill J, Dougan G, James KD, Thomson NR, Pickard D, Wain J, Churcher C, Mungall KL, Bentley SD, Holden MT, Sebaihia M, Baker S, Basham D, Brooks K, Chillingworth T, Connerton P, Cronin A, Davis P, Davies RM, Dowd L, White N, Farrar J, Feltwell T, Hamlin N, Haque A, Hien TT, Holroyd S, Jagels K, Krogh A, Larsen TS, Leather S, Moule S, O'Gaora P, Parry C, Quail M, Rutherford K, Simmonds M, Skelton J, Stevens K, Whitehead S, Barrell BG (2001) Complete genome sequence of a multiple drug resistant *Salmonella enterica* serovar Typhi CT18. Nature 413:848–852

Parry CM, Hien TT, Dougan G, White NJ, Farrar JJ (2002) Typhoid fever. New Engl J Med 347:1770–1782

Perry RD, Fetherston JD (1997) *Yersinia pestis*: the agent of the plague. Clin Microbiol Rev 10:35–66

Punjabi NH, Hoffman SL, Edman DC, Sukri N, Laughlin LW, Pulungsih SP, Rivai AR, Sututo, Moechtar A, Woodward TE (1988) Treatment of severe typhoid fever in children with high dose dexamethasone. Pediatr Infect Dis J 7:598–600

Raoult D, Aboudharam G, Crubézy E, Larroy G, Ludes B, Drancourt M (2000) Molecular identification by suicide PCR of *Yersinia pestis* as the agent of medieval black death. Proc Natl Acad Sci USA 23:12800–12803

Raoult D, Dutour O, Houhamdi L, Jankauskas R, Fournier PE, Ardagna Y, Drancourt M, Signoli M, La VD, Macia Y, Aboudharam G (2006) Evidence for louse-transmitted diseases in soldiers of Napoleon's Grand Army in Vilnius. J Infect Dis 193:112–120

Rogerson SJ, Spooner VJ, Smith TA, Richens J (1991) Hydrocortisone in chloramphenicol-treated severe typhoid fever in Papua New Guinea. Trans R Soc Trop Med Hyg 85:113–116

Scarrow GD (1988) The Athenian plague: a possible diagnosis. Ancient Hist Bull 2:4–8

Shapiro B, Rambaut A, Gilbert MTP (2006) No proof that typhoid caused the Plague of Athens (a reply to Papagrigorakis et al.). Int J Infect Dis 10:334–335

Shrewsbury JDF (1950) The plague of Athens. Bull Hist Med 24:1–24

Soupios MA (2004) Impact of the plague in Ancient Greece. Infect Dis Clin N Am 18:45–51

Stuart BM, Pullen RL (1946) Typhoid: clinical analysis of three hundred and sixty cases. Arch Intern Med 78:629–661

Taubenberger JK, Reid AH, Krafft AE, Bijwaard KE, Fanning TG (1997) Initial genetic characterisation of the 1918 Spanish influenza virus. Science 275:1793–1796

Taylor GM, Crossey M, Saldanha JA, Waldron T (1996) Detection of *Mycobacterium tuberculosis* bacterial DNA in medieval skeletal remains using polymerase chain reaction. J Archeol Sci 23:789–98

Taylor GM, Goyal M, Legge AJ, Shaw RJ, Young D (1999) Genotypic analysis of *Mycobacterium tuberculosis* from medieval human remains. Microbiology 145:899–904

Thucydides (1919) (Translation by C.F. Smith) History of the Peloponnesian War. Books I & II, 2.47–2.55. Harvard University Press, Cambridge

Chapter 11
Dental Pulp as a Tool for the Retrospective Diagnosis of Infectious Diseases

Vu Dang La(✉), Gerard Aboudharam, Didier Raoult, and Michel Drancourt

Abstract Dental pulp is a highly vascularised tissue of mesenchymal origin, located inside the tooth and naturally protected from the external environment. As with any soft tissue, dental pulp might contain microorganisms that circulate in the bloodstream. Therefore, dental pulp has been used as a tool for the detection of septicaemic infectious agents in both contemporary and ancient human and animal specimens. This chapter reviews the different methods used for the detection of microorganisms in dental pulp, most notably DNA-based methods. We propose a protocol for the use of dental pulp as a tool for the molecular detection of blood-borne microorganisms. This protocol would be useful for retrospective diagnosis of infectious diseases in palaeomicrobiology.

11.1 Introduction: What is Dental Pulp?

Dental pulp is a soft tissue of mesenchymal origin that occupies the central cavity and the root canals of teeth. The outermost layer in healthy pulp is the odontoblast layer of cells, which is located immediately below the predentin. The middle layer is a cell-poor zone in the coronal pulp that is sometimes not visible in young individuals; the inner layer is a cell-rich zone forming the dental pulp itself (Fig. 11.1). Dental pulp is a well vascularised tissue with arterioles entering the tooth through the apical foramina and accessory canals and passing centrally through the pulp, giving off lateral branches and dividing into capillaries. Smaller vessels reach the odontoblasts, where they divide extensively to form a plexus below and within the odontoblastic layer. Venous return is ensured by a network of capillaries that merge to form venules coursing down the central portion of the pulp. The vessels in dental pulp are of the terminal vascularisation type and have a density similar to

Vu Dang La
Unité des Rickettsies, CNRS UMR 6020, Faculté de Médecine, Université de la Méditerranée, 27 Boulevard Jean Moulin, 13385 Marseille Cedex 5, France
E-mail: danglv2001@yahoo.com

D. Raoult and M. Drancourt (eds.), *Paleomicrobiology: Past Human Infections.*

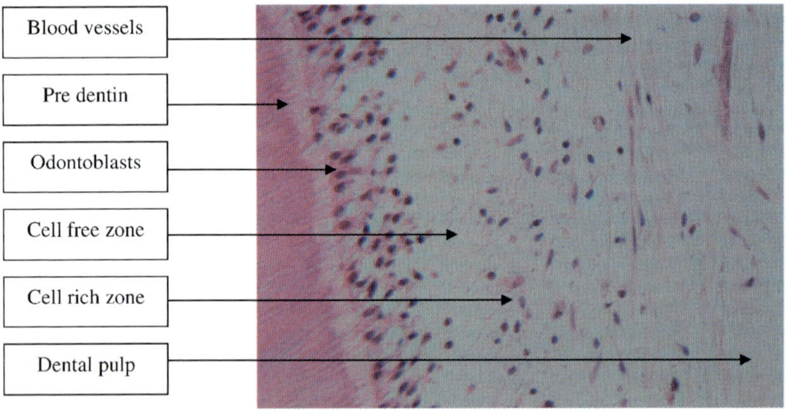

Fig. 11.1 Histology of mature human dental pulp. (Photograph: Dr. Marie Jos, Marseille Dental School, Marseille, France)

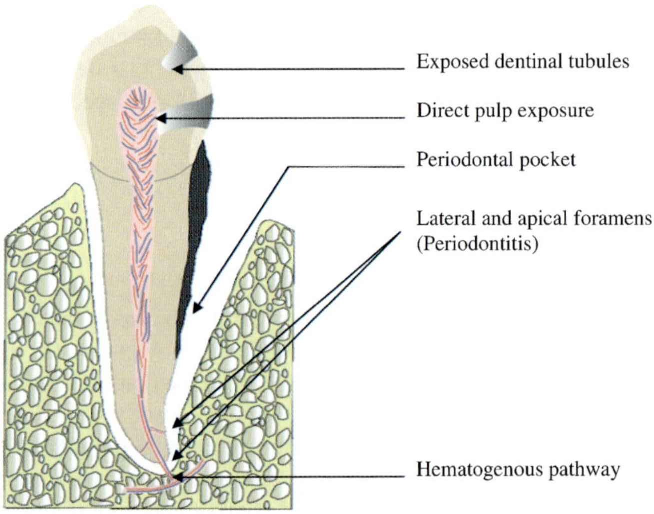

Fig. 11.2 Pathways of pulpal infection

that of vessels in the human brain (Kim 1985). Therefore, if an animal or a person has died due to a bacteraemia, it might be possible to find the pathogen responsible in the dental pulp. Changes that occur in dental pulp during life depend on the type of dentition growth. In rodents, there is continuous growth dentition, where hard dental tissues are continuously produced in order to replace those lost during chewing. The roots are open and there is usually a large volume of dental pulp. In other mammals, there is limited growth of dentition and, when teeth erupt, they are at

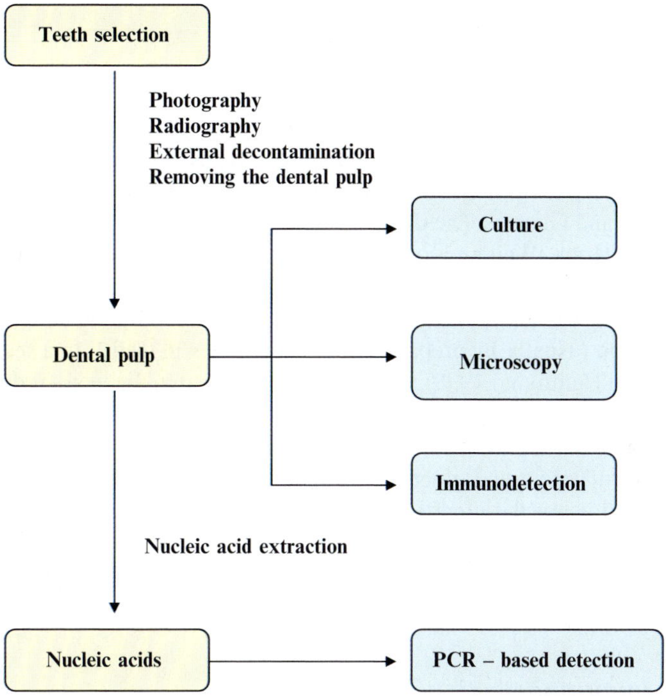

Fig. 11.3 Methods used for the detection of microorganisms in dental pulp

their final size, although the roots form completely only later. After the apical roots close, the pulp volume decreases gradually over time. Dentinogenesis is the inward formation of secondary dentin and is a continuous process. In people, continued growth of secondary dentin throughout life gradually reduces the size of the pulp chamber and the root canals. The dental pulp is protected from the environment by the surrounding dentin and enamel in the crown or cement in the root. Dental pulp can be infected by different routes, i.e. exposed dentinal, direct pulp exposure or via lateral or apical foramen and blood-borne microbes (Kettering 1994) (Fig. 11.2). This chapter focusses on the evidence for dental pulp infections resulting from blood-borne microorganisms and its application to the retrospective diagnosis of infectious diseases.

11.2 Methods Used for the Detection of Microorganisms in Dental Pulp

Methods used for the detection of microorganisms in dental pulp are summarised in Fig. 11.3.

11.2.1 Microscopy

The first study of bacterial colonisation of dental pulp in intact teeth was made by Tunnicliff and Hammond (1937). After disinfecting intact teeth and proving the sterility of their outside surfaces, dental pulp specimens were removed using sterile, fine-curved forceps and divided using sterile scissors. Smears were stained with gentian violet and Giemsa. The dental pulp was fixed in 1% formalin and sections stained with Gram–Weigert solution, hematoxylin-eosin and carbon thionin. Histology showed cocci and bacilli in different parts of the pulp without leukocyte infiltration. This was the first study of bacterial colonisation of dental pulp without dental irrigation (usually by preparing cervical cavities in teeth). In a second study (Tunnicliff and Hammond 1938), cavities were prepared in the teeth of dogs, which were intravenously injected with *Escherichia coli* and group A *Streptococcus*. After their removal, the teeth were fixed in 10% formalin or cold acetone, decalcified in 5% formic acid and embedded in paraffin. Six micrometric sections were cut and stained using the modified Brown and Brenn method (Brown and Brenn 1931). Bacteria were microscopically observed in the dental pulp of teeth removed at different times post-infection. Similar results were obtained by Tziafas (1989).

11.2.2 Immunodetection

The only immunodetection study of bacterial colonisation of dental pulp to date was performed by Gier and colleagues in 1968 in dogs (Gier and Mitchell 1968). Deep cavities were prepared in 88 teeth from four dogs, which were then inoculated intravenously with *E. coli*. A small piece of absorbent paper saturated with croton oil was sealed beneath a thick mixture of zinc oxide and eugenol in half of the cavities and the other half was left open to oral fluids. Bacteria were found in histological sections of unexposed pulp in 48 of 67 teeth in both prepared types of cavity and confirmed by culture. However, bacteria were also found in some blood vessels, but not in the pulpal tissue of one of the control teeth (unprepared) that was harvested 30 min after inoculation of bacteria. The authors remarked that the inflammatory reaction of the pulp was proportionate to the degree of injury and that bacteria were not found in 23 uninjured control teeth.

11.2.3 Culture

In a study in guinea pigs, Aboudharam et al. (2004a) prepared deep cervical cavities and injected bacteria intravenously. Of 15 unexposed dental pulps, 11 yielded group A *Streptococcus* and *E. coli* in culture. Similarly, culture of dental pulp yielded *Coxiella burnetii* in 2 of 12 samples after experimental bacteraemia in guinea pigs. No cavities were prepared and the cultures were positive at days 7 and

21 postinfection. These results suggested that dental pulp was equivalent to a small blood sample for the recovery of pathogenic agents. Viable *C. burnetii* were found in the dental pulp of guinea pigs after a week-long bacteraemia had ended (Aboudharam et al. 2004a). The authors suggested that dental pulp was a sanctuary for *C. burnetii* but the relevance of this finding to patients is unclear; guinea pigs have continuously developing dentition with open apexes, which is very different to the situation in human patients.

11.2.4 Nucleic-Acid-Based Detection

Nucleic-acid-based detection is the method most commonly used in this type of biomedical research. Nucleic acids extracted from dental pulp can be used for genomic amplification, most often by PCR, followed by sequencing and identification. This method has been used to make retrospective diagnoses of infectious diseases by detecting specific microbial sequences in dental pulp (Aboudharam et al. 2004a, 2004b, 2005; Drancourt et al. 1998, 2004, 2005; Glick et al. 1989; La et al. 2004, 2005; Papagrigorakis et al. 2006; Raoult et al. 2000; Wiechmann and Grupe 2005). In these studies, teeth were decontaminated, the dental pulp was removed by different methods, and PCR performed to search for specific fragments of microbial genomes. Such experiments have to be carefully controlled at each step because the method is highly sensitive and there is a high risk of contamination resulting in false positive results. We developed different protocols to minimise external contamination, including suicide PCR (Raoult et al. 2000) and carrying out experiments in laboratories where the bacteria of interest had never been introduced or worked on (Drancourt et al. 2004). We also modified a protocol initially proposed by Gilbert (Gilbert et al. 2003) whereby sterile auto-polymerisation resin was used to recover dental pulp without directly exposing it to the environment (La et al. 2005). All of these efforts (detailed in Sect. 11.3) helped to reduce the risks of contamination and to authenticate results. This approach is very useful in the study of ancient specimens where DNA has been shown to be better conserved in dental pulp than in bone (Ricaut et al. 2005), the most popular specimen used for the study of ancient DNA.

11.3 Detection of Infectious Agents in Dental Pulp During Bacteraemia

11.3.1 Experimental Models

Microbial colonisation of dental pulp was first studied in teeth with provoked inflammation. Early studies were performed in the 1940s–1960s and the target tissues in these studies were irrigated by preparing cervical cavities. The results, determined by microscopic or histopathological examination, did not distinguish infections arising from bacteraemia or from infections with oral bacteria because the dental pulp may be exposed via dentinal tubules when deep cavities are created in teeth. Delivanis further demonstrated that organisms did not appear in fluid collected following bacteraemia from the dental pulp of canine teeth that had no blood circulation (Delivanis et al. 1981). This phenomenon was recently confirmed by electron microscopy of dental pulp collected 1 day after a suspension of streptococci was injected intravenously in dogs (Tziafas 1989). These data indirectly suggested that, during bacteraemia, microorganisms can reach the dental pulp in cases of previous dental pulp inflammation. In these animal models, however, interpretation of the results is difficult because bacteria were identified in the pulp using only non-specific morphological criteria. It was later demonstrated that the dental pulp could be colonised by blood-borne bacteria in the absence of previous inflammation. Ten guinea pigs were inoculated intraperitoneally with *C. burnetii*, a strict intracellular bacterium responsible for Q fever that is not part of the normal flora of guinea pigs. At 20 days post-infection, in two out of four animals, the dental pulp was positive using PCR targeting two specific molecular fragments: primers CB1/CB2 targeting the gene encoding superoxide dismutase (*sod*) and primers Trans1/Trans2 targeting the insertion sequence IS111. Positive PCRs were found in 20–50% animals depending on the molecular target (Aboudharam et al. 2000). In this model, blood cultures were positive until the 5th day post-inoculation and spleen cultures were positive until the 10th day post-inoculation. *C. burnetii* DNA was not detected in the dental pulp until day 15 post-inoculation. These data showed that it was possible to detect specific DNA sequences in the dental pulp of bacteraemic animals. Moreover, detection was possible even after the bacteraemia ended. Further studies demonstrated that infection could also be demonstrated by direct culture of pulp tissue (Aboudharam et al. 2004a).

11.3.2 Naturally Infected Specimens

11.3.2.1 Animals

11.3.2.1.1 Cats and Bartonella

The domestic cat (*Felis silvestris catus*) is a reservoir for *Bartonella henselae*, *Bartonella clarridgeiae*, and *Bartonella koehlerae* (Droz et al. 1999; Kordick et al. 1997; Regnery et al. 1992). These species can cause human disease after a bite or scratch from cats or a cat flea (*Ctenocephalides felis*) bite (Chomel et al. 1996; Jacomo et al. 2002). In cats, *Bartonella* species cause chronic bacteraemia, which might persist for over a year without clinical or haematological changes (Abbot et al. 1997; Chomel et al. 2003; Koehler et al. 1994; Kordick et al. 1995). The prevalence of *B. henselae* bacteraemia in cats has varied from 4% to 68% in studies conducted in various countries worldwide (Koehler et al. 1994; Boulouis et al.

2005; Cholmel et al. 1995, 1999; Heller et al. 1997). The chronic asymptomatic bacteraemia that occurs with *Bartonella* spp. is an exceptional event in mammals that can help us to study the age and evolution of the relationships between cats and *Bartonella* spp. To develop a protocol for examining the remains of ancient cats, we investigated methods for the molecular detection of *Bartonella* spp. in 11 stray cats that had been buried for a year. We found that dental pulp was a suitable tissue for the molecular detection of *Bartonella* spp. and that dental pulp from the canine teeth of cats was statistically more likely to be positive by PCR detection than other teeth (Aboudharam et al. 2005). In a further study, we demonstrated that the dental pulp from 3 of 19 domestic cats dating from the thirteenth to the sixteenth centuries contained DNA specific to *B. henselae* (La et al. 2004). Also, DNA of *B. quintana* was found for the first time in contemporary domestic cats from Marseilles (La et al. 2005) even though humans were the only previously known reservoir for this bacterium (Maurin and Raoult 1996). This study led us to propose that cats might be an emerging source of human infections with *B. quintana*, in agreement with the epidemiological data in some cases of human *B. quintana* infections.

11.3.2.2 Humans

11.3.2.2.1 Humans and Human Immunodeficiency Virus

The first report of human immunodeficiency virus (HIV) in dental pulp was made by Glick, who used PCR to show HIV in a maxillary central incisor from a seropositive patient (Glick et al. 1989). The authors suggested that other viruses, such as hepatitis B, might also reside in the dental pulp, that instruments used for root canal therapy should be handled with the same caution as other sharp instruments, and that dental pulp should be disposed of in accordance with guidelines for other infected tissues. A more systematic study using PCR showed HIV in 11/12 pulps extracted from the teeth of 12 HIV seropositive patients. In situ hybridisation provided the first demonstration that HIV infects fibroblasts in the dental pulp. Histology did not reveal inflammation in the pulp but the authors suggested that dental pulp fibroblasts act as a reservoir for HIV in the body (Glick et al. 1991).

11.3.2.2.2 Humans and Herpes Simplex Virus

Herpes simplex virus (HSV) infects the oral cavity and migrates along the trigeminal nerve, part of which innervates the dental pulp. In a study of 46 patients, 19/23 of whom were seropositive (Heling et al. 2001), DNA of HSV could not be detected by PCR in dental pulp (11 normal and 17 necrotic), saliva or periapical tissue. The authors concluded that there is insufficient HSV for PCR detection, or that the virus did not enter the dental pulp. The authors did not specify how many seropositive patients were tested by PCR. Saliva from all the seropositive patients was tested by PCR and found negative although 7.4% of asymptomatic patients have viable HSV in oral rinse specimens of saliva (Spruance 1984).

11.3.2.2.3 Humans and Prion Protein

Based on animal model studies, Blanquet-Grossard et al. (2000) aimed to detect prions in the dental pulp of eight Creutzfeldt-Jakob patients using Western-blot analysis with monoclonal antibody 3F4, based on the results obtained in an animal model (Ingrosso et al. 1999). Although prions were found in the brains of patients, they were not detected in their dental pulp. The authors, however, had reservations about their results and recommended caution in health workers dealing with dental problems in patients with Creutzfeld-Jacob disease.

11.3.2.2.4 Humans and *Yersinia pestis*

The first demonstration of *Y. pestis* DNA in human dental pulp was made using 400-year-old samples (Drancourt et al. 1998). The results were confirmed using two different molecular targets (*pla* and *rpoB* genes) for *Y. pestis*. This study was also the first to provide nucleic-acid-based evidence of septicaemia in ancient remains in which there were no bone lesions indicative of the condition. Similar studies would be useful in resolving the etiology of other historical outbreaks, and the approach could be generally applied to research in palaeomicrobiology. In a further study we found DNA of *Y. pestis* in the dental pulp of victims of the medieval Black Death using a "suicide PCR" protocol. Here, the primers are used only once, there is no positive control, and positive specimens are sequenced and confirmed by sequencing a second target. This specific protocol minimises the risk of vertical contamination. Evidence that this technique was successful was our finding of original gene sequences that differed from sequences of modern strains; this protocol was proposed as the standard PCR technique of choice to completely avoid contamination of materials with previously amplified sequences (Raoult et al. 2000). Our results enabled us to resolve the long dispute over the aetiology of the Black Death by showing that the disease was in fact plague caused by *Y. pestis*. Independently, a German research group also detected *Y. pestis* DNA sequences in the teeth of two individuals buried in the second half of the sixth century A.D., which supported the evidence for the presence of *Y. pestis* in the first recorded pandemic (Weichmann and Grupe 2005). In this study, however, the DNA template was extracted from teeth powdered after several decontamination methods. In another study (Pusch et al. 2004), Yersinial F1 antigen was found in skeletons of victims of the Black Death, again validating the above result. However, an English research group (Gilbert et al. 2004) tried to detect *Y. pestis* in samples of specimens dating from the period of the Black Death using different molecular targets. This study did not find any specific fragment of *Y. pestis* DNA, and found DNA of a *Yersinia* strain only using a 16S fragment.

The new approach of genotyping bacteria from ancient specimens has been proposed as an appropriate method to study the epidemiology of ancient outbreaks of disease. Indeed, *Y. pestis* is subdivided into three major biovars that have been proposed to be responsible for the three specific plague pandemics. We collected dental pulp from individuals dating back to each of the two historical pandemics to test this hypothesis using the technique of multiple spacers typing (MST), which was developed in our laboratory (Drancourt et al. 2004). The results confirmed *Y. pestis* as the cause of the pandemics, and showed that the three pandemics were associated with the Orientalis biovar only (Drancourt et al. 2004).

11.3.2.2.5 Humans and *Salmonella enterica* serovar Typhi

The cause of the Plague of Athens has long been debated by scientists who have attempted to interpret Thucydides' descriptions of the disease. In 2006, dental pulp from remains in a mass burial pit dating from the outbreak (around 430 B.C.) was used to determine the probable cause of the Plague of Athens (Papagrigorakes et al. 2006). Although tests were performed, in random order, for several putative pathogens (*Yersinia pestis*, *Rickettsia prowazekii*, *Bacillus anthracis*, *Mycobacterium tuberculosis*, cowpox virus, *Bartonella henselae* and *Salmonella enterica* serovar Typhi) using previously developed protocols (Aboudharam et al. 2000, 2005; Drancourt et al. 1998, 2004; La et al. 2004) until a positive result was obtained, only DNA of *Salmonella enterica* serovar Typhi, using two molecular targets (*osmC*, *clyA* and *NarG*), was found in 3/3 of the teeth tested. In this study, the authors used dental pulp as the material of choice for retrospective diagnosis of bacteraemic agents and did not use dental powder (including dentine) since only dental pulp assures good vascularisation and durability, and is naturally protected from external contamination. The protocol enables dental pulp that has been protected inside teeth for centuries to be used as the only available equivalent to a blood sample in the diagnosis of an infection. Extreme measures were taken to prevent any possibility of exogenous contamination, including suicide PCR, targeting different genomic regions and blinded manipulations by different operators in several laboratories. The positive results revealed original sequences that were repeatedly obtained by independent operators, thus validating the results. Once again, dental pulp proved useful in providing clear evidence leading to retrospective diagnoses from ancient remains and helped determine the cause of the Antique Plague of Athens.

11.3.2.2.6 Humans and *Bartonella quintana*

B. quintana is the etiological agent of trench fever, which occurred in soldiers during World Wars I and II (Byam et al. 1919; Kostrzewski 1949). This bacterium has now also been reported to cause chronic bacteraemia and endocarditis in homeless and alcoholic patients in modern cities in both Europe and the United States (Drancourt et al. 1995; Brouqui et al. 1999; Spach et al. 1995; Stein and Raoult 1995) and bacillary angiomatosis in both HIV-infected and immunocompromised patients (Koehler et al. 1997). Using PCR and sequencing, *B. quintana* DNA has been detected in the dental pulp of a patient who had been successfully treated with antibiotics for *B. quintana* septicaemia 6 months previously (Aboudharam et al. 2004b). Blood cultures were negative at the time the dental pulp was found positive, therefore suggesting that dental pulp was a sanctuary for *B. quintana* DNA. We have also shown that *B. quintana* can be detected in dental pulp from the remains of a person who died 4,000 years ago; this study was the first to demonstrate *B. quintana* DNA in ancient human remains (Drancourt et al. 2005). A recent study of dental pulp from the remains of Napoleon's soldiers showed that louse-borne infectious diseases caused by *B. quintana* and *Rickettsia prowazekii* affected nearly one-third of Napoleon's soldiers buried in Vilnius and might have been a major factor in the French retreat from Russia (Raoult et al. 2006). This study once again showed that dental pulp can be used for the retrospective diagnosis of infectious diseases.

11.3.2.2.7 Humans and *Rickettsia prowazekii*

Rickettsia prowazekii causes epidemic typhus in people during wartime and has been classified on the B list of potential bioterrorism agents by the Centers for Disease Control and Prevention (Atlanta, GA). The bacterium was found in the dental pulp of three soldiers from Napoleon's army (Raoult et al. 2006) and, since epidemic typhus results in high mortality, it was thought likely that the soldiers died of the disease. This study again confirms that searching for the DNA of infectious agent in dental pulp is an important tool in investigating the history of infectious diseases (Raoult et al. 2006).

11.4 Protocols for Molecular Detection of Microorganisms in Dental Pulp

11.4.1 Selection and Preparation of the Teeth

The selection of teeth is the first step (Box 11.1). To be suitable, a tooth must be intact, with a closed apex that will prevent external contamination – the most

Box 11.1 Suggested guidelines for the selection of teeth prior to total DNA extraction from dental pulp

For microbial detection, teeth should be transported and stored in separate containers at room temperature with the following stipulations:

Teeth should be intact with a closed apex

Teeth with single roots are preferred

Unerupted teeth should be tested immediately after being exposed to the external environment

More than one tooth per individual should be used if possible important factor when working with ancient specimens. The presence of a closed apex should be reconfirmed after the tooth has been washed and subjected to external decontamination as it is sometimes very difficult to observe tiny defects because of the colour and nature of ancient specimens. In our experience, teeth with a single root are more suitable for study because they have a large volume of pulp and are easier to manipulate than teeth with multiple roots. The canine tooth of people and cats is particularly suitable because it has the largest volume of pulp. Unerupted teeth are totally protected in the jawbone but their apex is almost always open. Therefore, to prevent contamination, any experiment on unerupted teeth should be performed immediately after removing the tooth. The quality of ancient samples cannot be assured in all specimens, and, if possible, multiple teeth should be examined per individual. Once a tooth has been selected, digital photographs and radiographs are taken for identification purposes, and to record information on the specimens such as their form and colour. Radiographs enable operators to estimate the pulp volume, the presence of calcification in the pulp and the status of the apex. This is important as calcification of pulp and very small pulp volume, although rarely encountered, prevent the recovery of sufficient pulp material for proper detection of microorganisms. Our experience with a large collection of human teeth has indicated that the presence of dental pulp calcification in teeth makes the pulp unsuitable for nucleic acid extraction. Digital information (photographs and X-ray radiographs) is also useful for subsequent anthropological studies because the tooth suffers considerable damage when dental pulp is extracted. However, we have been able to reassemble and glue teeth back together after dental pulp removal and thus restore some of their initial appearance.

11.4.2 Removal of Dental Pulp

Dental pulp can be removed using different methods depending on its status and the purpose of the study (Figs. 11.4, 11.5, Box 11.2). Generally, teeth are extracted and decontaminated by wiping their external surface with bleach and by exposure to ultraviolet (UV) light before the dental pulp is removed. In one method, the entire tooth is crushed into a fine powder, decalcified, and nucleic acids then extracted. In a second method, which is applied to teeth from living people, the tooth is cleansed, isolated with a rubber dam, opened with a sterile bur and the dental pulp removed with a sterile broach (Glick et al. 1989, 1991; Heling et al. 2001). This method is rarely appropriate because we always try to keep dental pulp as alive as possible in the patient. This protocol should be used only in cases of irreversible pulpitis or other therapeutic indication for dental pulp removal. A third method consists of opening the teeth from ancient remains with a longitudinal fracture using a rotative disk, and scraping the dental pulp and its powdery remnants into a sterile tube for DNA extraction (Aboudharam et al. 2004a, 2005; Drancourt et al. 1998, 2004; La et al. 2004; Papagrigorakis et al. 2006; Raoult et al. 2006). This method is simple to carry out. All the above methods carry some risk of contamination because

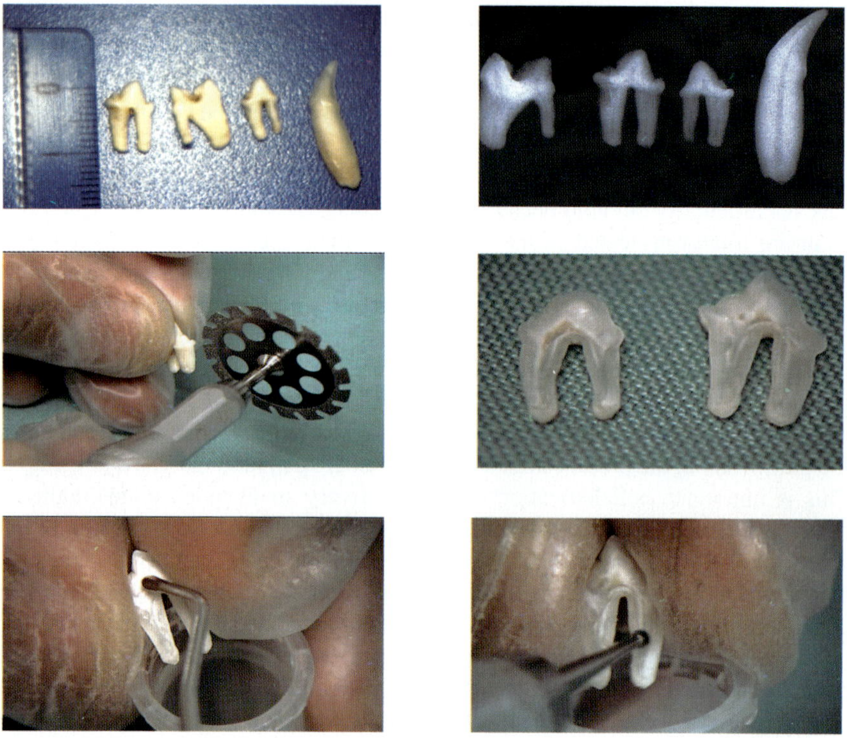

Fig. 11.4 Protocol for recovery of dental pulp from feline teeth

dental pulp is exposed to the environment while being removed. An original protocol for recovery of dental pulp via the apex was established by Gilbert et al. (2003). Here, the tooth is fully encased in silicone rubber, the top of the root is removed horizontally and the dental pulp is powdered and removed from the pulp chamber using a dental drill bit. In our experience, however, the silicone does not tightly adhere to the tooth; we tried to improve the protocol by firstly decontaminating the tooth with 70% ethanol and then placing the entire tooth in sterile resin (Resin Polyester SODY 33, ESCIL, Chassieu, France) poured into a sterile centrifugation tube (Millipore, Bedford, MA). After polymerisation of the resin at room temperature for 30 min, the apex of the root is removed using a sterile disk and the opened tooth, still embedded in the resin, is inserted upside down in a sterile Eppendorf tube and centrifuged at 8,000 rpm for 10 min to recover the dental pulp (La et al. 2005). This method has been applied to the recovery of dental pulp from contemporary teeth but not yet from ancient teeth. In ancient teeth, the pulpal chamber is not entirely occupied by dried dental pulp remnants, thus DNA extraction reagents can be injected directly into these cavities via the apex. This protocol uses the tooth itself as a sterile tube to contain the reagents but it requires a large pulp chamber.

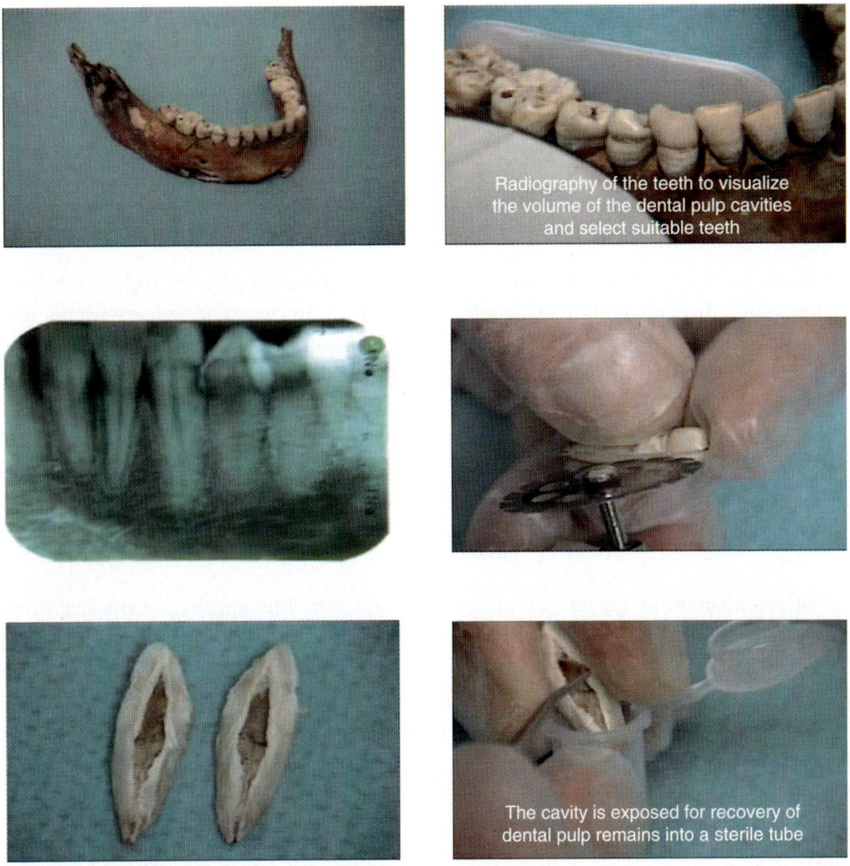

Fig. 11.5 Protocol for recovery of dental pulp from human teeth

Box 11.2 Suggested protocol for the recovery of dental pulp

Photography
Radiography
External decontamination
  * 70% Ethanol
  * Ultra violet (for immunohistology)
Enclose tooth with sterile resin using a type centrifugation tube
Moving top of the tooth to open root canal widely enough
With contemporary teeth, centrifuge the pulp into a collection tube (10 min at 8,000 rpm)
With ancient teeth, inject extraction reagents (Proteinase K, SDS) and incubate according to extraction protocol prior to the above step

For that reason, continued dentine formation results in a reduced volume of the dental pulp chamber, which precludes the use of this method (L. Tran-Hung, personal communication).

11.4.3 Extraction of Nucleic Acids

Dental pulp is a soft tissue and any extraction protocol can be used to extract nucleic acids directly without a decalcification step. In our experience with ancient specimens, the phenol–chloroform protocol for total nucleic acid extraction is the optimal protocol. Controls are essential to detect contamination but, to avoid the risk of contamination, positive controls should not be used.

11.4.4 Genomic Amplification

We select molecular targets of under 300 bp, carry out amplification reactions in small volumes (final volume: 25 μl), and add BSA (bovine serum albumin) to our reaction mixture to reduce the influence of inhibitory factors present in ancient samples. We have not used positive controls in our PCR-based experiments, preferring to use a "suicide PCR" protocol to minimise the risk of contamination.

11.4.5 Prevention of Contamination

This is the most important aspect of the procedure and, depending on the nature of the sample, we use a variety of methods for external decontamination that include cleaning teeth with distilled water, immersing them in absolute or 70% ethanol, and/or exposing them to UV light. The best protocol is to totally cover the tooth with sterile resin and then recover dental pulp through the apex by centrifugation. This technique, however, is not suitable for molars with very narrow root canals. It is most useful for teeth with single roots.

Every step in the experiment should be performed in a separate room with disposable equipment and newly prepared reagents. All PCR-based experiments should be carried out in designated one-way PCR suites with appropriate ventilation. "Suicide PCR" reactions that target a new genomic region may prevent vertical contamination from previous amplifications. The introduction of numerous contamination and negative controls in any amplification reaction may help to ascertain the source of any contamination. In our laboratory, we now use one negative control and one contamination control for every four ancient samples we process.

11.5 Applications in Palaeomicrobiology

Dental pulp has been used for the detection, identification and characterisation of microorganisms in ancient remains using PCR-based molecular techniques (Drancourt et al. 1998, 2004, 2005; La et al. 2004; Papagrigorakis et al. 2006; Raoult et al. 2000, 2006; Weichmann and Grupe 2005).The consecutive publication of several papers implementing the more or less same experimental protocol underlies the increasing worldwide acceptance of its validity and applicability on several relevant matters of historical and medical interest. Thus, investigating dental pulp has allowed us to suggest certain bacteria as the possible etiologic causes involved in the host's bacteraemia, and has helped interpret historical and anthropologic data in which there was no microbial evidence and where it is hard to find exact descriptions in comparison to modern pathology due to ancient language and problems of translation. These studies strongly encourage other researchers who are interested in investigating the history of infectious diseases to use dental pulp as the material of choice for their research. Recently, by using dental pulp, we clearly demonstrated that louse-borne infectious diseases affected nearly one-third of Napoleon's soldiers buried in Vulnius, and indicated that these diseases might have been a major factor in the French retreat from Russia (Raoult et al. 2006). Similarly, a Greek team led by Manolis Papagrigorakis successfully applied this technique to shed light on one of the most debated enigmas in medical history, the cause of the Plague of Athens (Papagrigorakis et al. 2006); this report pinpoints typhoid fever as the disease responsible for this devastating epidemic. Hence, investigation of dental pulp allows us to diagnose past infectious diseases and elucidate past epidemiologies by determining the causative organism. However, a fundamental problem is the need for careful measures to protect material from external contamination. In order to validate the data, we proposed a set of criteria that can be used to assess results obtained from ancient specimens (Box 11.3). Because of its durability, using dental pulp to detect microorganisms might help us determine the time during which the microbes infected the host. Dental pulp can also be used to provide information on emerging infectious diseases by helping to establish models of emerging infections and by contributing to the development of appropriate preventive measures (Drancourt and Raoult 2005). In fact, we unexpectedly discovered that cats could be infected by presumably bacteraemic *B. quintana* (La et al. 2005). This observation is very useful in understanding the natural epidemiologic cycle of *B. quintana*, from which we can recommend that immunocompromised patients avoid contacts with cats. For the first time, we were able to demonstrate this missing link in *B. quintana* infections by using dental pulp. Therefore, we can apply this approach to identify *Bartonella* spp. in other animals and locations, especially in ancient animals; this will help in the understanding of the geographical distribution of this bacterial genus. We have been able to show that the co-evolution between *B. henselae* genotype Houston and cats existed at least 800 years ago in France, even though the first description of cat scratch disease (CSD) caused by *B. henselae* dates from 1950 (Debré et al. 1950). The high prevalence of *Bartonella* spp.

Box 11.3 Criteria for the authentication of molecular data in palaeomicrobiology (Drancourt and Raoult 2005) (reproduced with permission from Nature Reviews)

Absence of a positive control
The positive control should be removed from the laboratory in which ancient specimens are processed
Negativity of negative controls
Several negative controls should be analysed in parallel with the specimens being processed
Negative controls should be as similar as possible to the ancient specimens
Negative controls should remain free of amplicons
Sequencing of PCR amplicons
PCR alone does not ensure the specificity of the diagnosis, and amplicons have to be sequenced to identify ancient microorganisms
Targetting a new sequence in the laboratory
PCR should target a specific sequence that has not previously been amplified in the laboratory
Amplification and sequencing of a second target
A positive result must be confirmed by amplification and sequencing of a second specific molecular target
Originality of the ancient sequence
Acquisition of an original sequence that differs from modern homologues by mutation or deletion excludes contamination

bacteraemia in cats supports the possibilities of frequent exposure, persistent infection, and recurrent infection with this bacterium in cats. Therefore, it can be assumed that cats have long been infected by this bacterium and that the two have co-evolved over centuries. This may provide very reasonable explanations of some historical medical sources, such as reports of the French and English kings' apparent power to cure scrofula in medieval times – some of the cases may have been self-limiting CSD (La et al. 2004). The antiquity of this co-evolution, in which cats and *B. henselae* have interacted over many centuries, could be further studied by increasing the number of specimens as well as the time frame they cover. It would also be very interesting to enlarge the number of ancient cat samples from different countries, such as those of the New World, in order to study the genomic variations and the origin of this bacterium. Indeed, this seems very feasible because we have already demonstrated that dental pulp is useful not only for detection of infectious diseases but also for genotyping ancient bacteria (cf. our *Y. pestis* study, see above) and such studies have the potential to contribute greatly to genomic research (Drancourt et al. 2004; Drancourt and Raoult 2005). We recommend that dental pulp of ancient remains be used for retrospective diagnoses; pathogens found in dental pulp might have been associated with bacteraemia in the host. We would note that viable bacteria can be found in dental pulp even though blood cultures are negative or unavailable (Aboudharam 2004a, 2004b). Therefore, we can use dental pulp to search for blood-borne microorganisms in cases where blood tests are not possible, as in ancient specimens and, in certain cases, investigation of dental pulp may provide evidence to support putative historical hypotheses (Papagrigorakis et al 2006; Drancourt and Raoult 2005). Because targeting a specific pathogen has its limitations in molecular diagnosis, the 16S rRNA gene has been used as a universal detection fragment for bacteria. However, this approach is prone to problems of contamination. The only attempted 16S rRNA gene-based detection of bacteria in ancient dental pulp to date resulted in contaminated amplifications. Therefore, a more universally applicable protocol will be required to promote the use of dental pulp in palaeomicrobiology.

11.6 Conclusions

Bacteria can colonise dental pulp via the haematogenous route, and their presence can be demonstrated in this tissue by using molecular techniques in both contemporary and ancient specimens, as well as in culture-based experiments for contemporary specimens. Further investigations are required to determine the range of bacteria that can colonise dental pulp from the blood. In some infections, dental pulp is considered as a sanctuary for microorganisms, which can be detected by PCR even after blood cultures become negative. This might be because these microorganisms are no longer in the bloodstream but can reside in the dental pulp; this tissue can then be used to detect blood-borne pathogens even if blood cultures are not available or are found to be negative. A summary of published data (Table 11.1) shows that our laboratory has contributed greatly to this field of research since 1998 by developing and improving the techniques required. From our experiences and the available literature we would note that there is an increased chance of finding bacteria in teeth that have a large volume of dental pulp, e.g. the canine teeth of cats and humans, as compared to other teeth. It is easier to obtain sequences using contemporary teeth, as positive samples can be sequenced directly. With ancient teeth, however, molecular cloning is usually necessary to determine sequences from positive samples. Prevention of contamination is essential, as ancient samples are unique and there are few available materials to use for further investigations.

Table 11.1 Summary of published studies of detection of infectious agents in dental pulp. *PCR* Polymerase chain reaction, *HIV* human immunodeficiency virus, *HSV* herpes simplex virus

Source	Date	Infection method	Method	Number of tested teeth	PCR target	Microorganism	Reference
Dog	Modern	Intravenous with irritation	Culture / histology / immunohistology	109		*Escherichia coli* beta-hemolytic *Streptococcus*	Aboudharam et al. 2004a
Dog	Modern	Intravenous with irritation	Histology	36		*Streptococcus* spp.	Aboudharam et al. 2004b
Cat	Thirteenth–sixteenth century	Natural colonisation	PCR	135	*groEL*; *Pap31*	*Bartonella henselae*	Papagrigorakis et al. 2006[a]
Cat	Modern	Natural colonisation	PCR	9	*ITS*; *Pap31*	*Bartonella henselae*; *Bartonella quintana*	Raoult et al. 2000[a]
Cat	Mimic ancient	Natural colonisation	PCR	104	*groEL*	*Bartonella henselae*; *Bartonella* spp.	Drancourt et al. 2004[a]
Guinea-pig	Modern	Intraperitoneal without irritation	PCR	280	*Sod*; *IS111*	*Coxiella burnetii*	Regnery et al. 1992[a]
Guinea-pig	Modern	Intraperitoneal without irritation	Culture	52		*Coxiella burnetii*	Aboudharam et al. 2005[a]
Human	Modern	Natural colonisation	Culture / histology	30		*Streptococci*	Tziafas 1989
Human	Modern	Natural colonisation	PCR	1	Antibodies to HIV	HIV	La et al. 2005
Human	Modern	Natural colonisation	PCR / hybridisation	12	Antibodies to HIV	HIV	Blanquet-Grossard et al. 2000
Human	Modern	Natural colonisation	Western blot	8		Prion protein	Gilbert et al. 2004
Human	1348 A.D.	Natural colonisation	PCR	23	*pla*	*Yersinia pestis*	Ricaut et al. 2005[a]
Human	Sixteenth–eighteenth century	Natural colonisation	PCR	12	*pla*; *rpoB*	*Yersinia pestis*	Drancourt et al. 2005[a]

Human	Fifth–fourteenth century	Natural colonisation	PCR	19	Multiple intergenic spacers	*Yersinia pestis*	Glick et al. 1989[a]
Human	Modern	Natural colonisation	PCR	1	*groEL*; *hppE*	*Bartonella quintana*	Drancourt et al. 1998[a]
Human	2000 B.C.	Natural colonisation	PCR	12	*groEL*; *hppE*	*Bartonella quintana*	La et al. 2004[a]
Human	Sixth century	Natural colonisation	PCR	2	*pla*	*Yersinia pestis*	Delivanis et al. 1981
Human	Modern	Natural colonisation	PCR	28	Quanti-PATH HSV 1,2 KIT	Herpes simplex virus	Ingrosso et al. 1999
Human	Thirteenth–seventeenth century	Natural colonisation	PCR	108	16S; *pla*	*Yersinia pestis*	Aboudharam et al. 2000
Human	Eighteenth century	Natural colonisation	PCR	47	*glpD*	*Yersinia pestis*	Drancourt et al. 2007[a]
Human	1812	Natural colonisation	PCR	86	16S; *pla*	*Bartonella quintana*; *Rickettsia prowazekii*	Gilbert et al. 2003[a]
Human	430 B.C.	Natural colonisation	PCR	3	*OsmC Et clyA* and *narG*	*Salmonella entica* (serovar *Typhi*)	Wiechmann and Grupe 2005
Human	Modern	Natural colonisation	PCR	51	16S; and *rpoB*	All bacteria	L. Tran-Hung et al. 2007

[a] Studies performed in our laboratory - 13/23 (56 %)

Acknowledgements The authors acknowledge Dr. Patrick Kelly for expert reviewing of the manuscript.

References

Abbott RC, Chomel BB, Kasten RW, Floyd-Hawkins KA, Kikuchi Y, Koehler JE, Pedersen NC (1997) Experimental and natural infection with *Bartonella henselae* in domestic cats. Comp Immunol Microbiol Infect Dis 20:41–51

Aboudharam G, Lascola B, Raoult D, Drancourt M (2000) Detection of *Coxiella burnetii* DNA in dental pulp during experimental bacteraemia. Microb Pathog 28:249–254

Aboudharam G, Fournier PE, Drancourt M, Raoult D, Foucault C, Brouqui P (2004a) Molecular detection of *Bartonella quintana* DNA in the dental pulp of a homeless patient. Eur J Clin Microbiol Infect Dis 23:920–922

Aboudharam G, Drancourt M, Raoult D (2004b) Culture of *C. burnetii* from the dental pulp of experimentally infected guinea pigs. Microb Pathog 36:349–350

Aboudharam G, Vu DL, Davoust B, Drancourt M, Raoult D (2005) Molecular detection of *Bartonella* spp. in the dental pulp of stray cats buried for a year. Microb Pathog 38:47–51

Blanquet-Grossard F, Sazdovitch V, Jean A, Deslys JP, Dormont D, Hauw JJ, Marion D, Brown P, Cesbron JY (2000) Prion protein is not detectable in dental pulp from patients with Creutzfeldt-Jakob disease. J Dent Res 79:700

Boulouis HJ, Chang CC, Henn JB, Kasten RW, Chomel BB (2005) Factors associated with the rapid emergence of zoonotic Bartonella infections. Vet Res 36:383–410

Brouqui P, Lascola B, Roux V, Raoult D (1999) Chronic *Bartonella quintana* bacteraemia in homeless patients. N Engl J Med 340:184–189

Brown JH, Brenn L (1931) Bull Johns Hopkins Hosp 48:69 [cited in Lina G (ed) AFIP Manual of Histologic Staining Techniques, 3rd edn, McGraw-Hill, New York, p 222]

Byam W, Caroll JH, Churchill JH, Dimond L, Sorapure VE, Wilson RM, Lloyd LL (1919) Trench fever: a louse-born disease. Frowde and Hodder Stoughton, London

Chomel BB, Abbott RC, Kasten RW, Floyd-Hawkins KA, Kass PH, Glaser CA, Pedersen NC, Koehler JE (1995) *Bartonella henselae* prevalence in domestic cats in California: risk factors and association between bacteraemia and antibody titers. J Clin Microbiol 33:2445–2450

Chomel BB, Kasten RW, Floyd-Hawkins K, Chi B, Yamamoto K, Roberts-Wilson J, Gurfield AN, Abbott RC, Pedersen NC, Koehler JE (1996) Experimental transmission of *Bartonella henselae* by the cat flea. J Clin Microbiol 34:1952–1926

Chomel BB, Carlos ET, Kasten RW, Yamamoto K, Chang CC, Carlos RS, Abenes MV, Pajares CM (1999) *Bartonella henselae* and *Bartonella clarridgeiae* infection in domestic cats from The Philippines. Am J Trop Med Hyg 60:593–597

Chomel BB, Kasten RW, Sykes JE, Boulouis HJ, Breitschwerdt EB (2003) Clinical impact of persistent Bartonella bacteraemia in humans and animals. Ann N Y Acad Sci 990:267–278

Debré R, Lamy M, Jammet ML, Costil L, Mozziconacci P (1950) La maladie des griffes du chat. Soc Med Hôp Paris 66:76–79

Delivanis PD, Snowden RB, Doyle RJ (1981) Localization of blood-borne bacteria in instrumented unfilled root canals. Oral Surg Oral Med Oral Pathol 52:430–432

Drancourt M, Raoult D (2005) Palaeomicrobiology: current issues and perspectives. Nat Rev Microbiol 3:23–35

Drancourt M, Mainardi JL, Brouqui P, Vandenesch F, Carta A, Lehnert F, Etienne J, Goldstein F, Acar J, Raoult D (1995) *Bartonella* (*Rochalimaea*) *quintana* endocarditis in three homeless men. N Engl J Med 332:419–423

Drancourt M, Aboudharam G, Signoli M, Dutour O, Raoult D (1998) Detection of 400-year-old *Yersinia pestis* DNA in human dental pulp: an approach to the diagnosis of ancient septicemia. Proc Natl Acad Sci USA 95:12637–12640

Drancourt M, Roux V, Dang LV, Tran-Hung L, Castex D, Chenal-Francisque V, Ogata H, Fournier PE, Crubézy E, Raoult D (2004) Genotyping, Orientalis-like *Yersinia pestis*, and plague pandemics. Emerg Infect Dis 10:1585–1592

Drancourt M, Tran-Hung L, Courtin J, Lumley H, Raoult D (2005) *Bartonella quintana* in a 4000-year-old human tooth. J Infect Dis 191:607–611

Drancourt M, Signoli M, Dang LV, Bizot B, Roux V, Tzortzis S, Raoult D (2007) *Yersinia pestis* Orientalis in remains of ancient plague patients. Emerg Infect Dis 13:332–333

Droz S, Chi B, Horn E, Steigerwalt AG, Whitney AM, Brenner DJ (1999) *Bartonella koehlerae* sp. nov., isolated from cats. J Clin Microbiol 37:1117–1122

Gier RE, Mitchell DF (1968) Anachoretic effect of pulpitis. J Dent Res 47:564–570

Gilbert MT, Willerslev E, Hansen AJ, Barnes I, Rudbeck L, Lynnerup N, Cooper A (2003) Distribution patterns of postmortem damage in human mitochondrial DNA. Am J Hum Genet 72:32–47

Gilbert MT, Cuccui J, White W, Lynnerup N, Titball RW, Cooper A, Prentice MB (2004) Absence of *Yersinia pestis*-specific DNA in human teeth from five European excavations of putative plague victims. Microbiology 150:263–265

Glick M, Trope M, Pliskin E (1989) Detection of HIV in the dental pulp of a patient with AIDS. J Am Dent Assoc 119:649–650

Glick M, Trope M, Pliskin E (1991) Human immunodeficiency virus infection of fibroblasts of dental pulp in seropositive patients. Oral Surg Oral Med Oral Pathol 71:733–736

Heling I, Morag-Hezroni M, Marva E, Hochman N, Zakay-Rones Z, Morag A (2001) Is herpes simplex virus associated with pulp/periapical inflammation? Oral Surg Oral Med Oral Pathol Oral Radiol Endod 91:359–361

Heller R, Artois M, Xemar V, De Briel D, Gehin H, Jaulhac B, Monteil H, Piemont Y (1997) Prevalence of *Bartonella henselae* and *Bartonella clarridgeiae* in stray cats. J Clin Microbiol 35:1327–1331

Ingrosso L, Pisani F, Pocchiari M (1999) Transmission of the 263K scrapie strain by the dental route. J Gen Virol 80:3043–3047

Jacomo V, Kelly PJ, Raoult D (2002) Natural history of Bartonella infections (an exception to Koch's postulate). Clin Diagn Lab Immunol 9:8–18

Kettering JD, Torabinejad M (1994) Microbiology and immunology. In: Cohen C, Burns E (eds) Pathways of the dental pulp, 6th edn. Mosby, St. Louis, pp 363–376

Kim S (1985) Regulation of pulpal blood flow. J Dent Res 64:590–596

Koehler JE, Glaser CA, Tappero JW (1994) *Rochalimaea henselae* infection. A new zoonosis with the domestic cat as reservoir. JAMA 271:531–535

Koehler JE, Sanchez MA, Garrido CS, Whitfeld MJ, Chen FM, Berger TG, Rodriguez-Barradas MC, LeBoit PE, Tappero JW (1997) Molecular epidemiology of Bartonella infections in patients with bacillary angiomatosis-peliosis. N Engl J Med 337:1876–1883

Kordick DL, Wilson KH, Sexton DJ, Hadfield TL, Berkhoff HA, Breitschwerdt EB (1995) Prolonged Bartonella bacteraemia in cats associated with cat-scratch disease patients. J Clin Microbiol 33:3245–3251

Kordick DL, Hilyard EJ, Hadfield TL, Wilson KH, Steigerwalt AG, Brenner DJ, Breitschwerdt EB (1997) *Bartonella clarridgeiae*, a newly recognized zoonotic pathogen causing inoculation papules, fever, and lymphadenopathy (cat scratch disease). J Clin Microbiol 35:1813–1818

Kostrzewski J (1949) The epidemiology of trench fever. Bull Acad Pol Sci Med 7:233–263

La VD, Clavel B, Lepetz S, Aboudharam G, Raoult D, Drancourt M (2004) Molecular detection of *Bartonella henselae* DNA in the dental pulp of 800-year-old French cats. Clin Infect Dis 39:1391–1394

La VD, Tran-Hung L, Aboudharam G, Raoult D, Drancourt M (2005) *Bartonella quintana* in domestic cat. Emerg Infect Dis 11:1287–1289

Maurin M, Raoult D (2006) *Bartonella* (*Rochalimaea*) *quintana* infections. Clin Microbiol Rev 9:273–292

Papagrigorakis MJ, Yapijakis C, Synodinos PN, Baziotopoulou-Valavani E (2006) DNA examination of ancient dental pulp incriminates typhoid fever as a probable cause of the Plague of Athens. Int J Infect Dis 10:335–336

Pusch CM, Rahalison L, Blin N, Nicholson GJ, Czarnetzki A (2004) Yersinial F1 antigen and the cause of Black Death. Lancet Infect Dis 4:484–485

Raoult D, Aboudharam G, Crubézy E, Larrouy G, Ludes B, Drancourt M (2000) Molecular identification by "suicide PCR" of *Yersinia pestis* as the agent of medieval black death. Proc Natl Acad Sci USA 97:12800–12803

Raoult D, Dutour O, Houhamdi L, Jankauskas R, Fournier PE, Ardagna Y, Drancourt M, Signoli M, La VD, Macia Y, Aboudharam G (2006) Evidence for louse-transmitted diseases in soldiers of Napoleon's grand army in Vilnius. J Infect Dis 193:112–120

Regnery R, Martin M, Olson J (1992) Naturally occurring "*Rochalimaea henselae*" infection in domestic cat. Lancet 34:557–558

Ricaut FX, Keyser-Tracqui C, Crubézy E, Ludes B (2005) STR-genotyping from human medieval tooth and bone samples. Forensic Sci Int 151:31–35

Spach DH, Kanter AS, Dougherty MJ, Larson AM, Coyle MB, Brenner DJ, Swaminathan B, Matar GM, Welch DF, Root RK, Stamm WE (1995) *Bartonella* (*Rochalimaea*) *quintana* bacteraemia in inner-city patients with chronic alcoholism. N Engl J Med 332:424–428

Spruance SL (1984) Pathogenesis of herpes simplex labialis: excretion of virus in the oral cavity. J Clin Microbiol 19:675–679

Stein A, Raoult D (1995) Return of trench fever. Lancet 345:450–451

Tran-Hung L, Tran-Thi N, Aboudharam G, Raoult D, Drancourt M (2007) A new method to extract dental pulp DNA: application to universal detection of bacteria. PLoS ONE 10:e1062

Tunnicliff R, Hammond J (1937) Presence of bacteria in pulps of intact teeth. J Am Dental Assoc 24:1663–1666

Tunnicliff R, Hammond C (1938) Smooth and rough greening streptococci in pulps of intact and various teeth and in various dentin. J Am Dental Assoc 25:1046–1052

Tziafas D (1989) Experimental bacterial anachoresis in dog dental pulps capped with calcium hydroxide. J Endod 15:591–595

Wiechmann I, Grupe G (2005) Detection of *Yersinia pestis* DNA in two early medieval skeletal finds from Aschheim (Upper Bavaria, 6th century A.D.). Am J Phys Anthropol 126:48–55

Viruses

Chapter 12
History of Influenza Pandemics

Bruno Lina

Abstract Influenza pandemics have been amongst the largest and the deadliest epidemics in the history of man, and were observed already in ancient times. For example, records from the fifth century B.C. suggest that influenza pandemics were observed in ancient Greece. In Europe, during the Middle Ages and the Renaissance, numerous concordant reports from different countries describe epidemics of respiratory infections that resemble influenza pandemics. However, it is not possible to be certain that these epidemics were due to influenza. In the twentieth century, three influenza pandemics have occurred, including the deadly Spanish flu pandemic. Modern virology has unravelled the mechanisms of emergence of pandemic viruses, and considerable knowledge on influenza viruses has been accumulated. The picture is now clear: influenza A is a zoonotic virus whose reservoir is in wild birds. In rare cases, these avian viruses are introduced into man and, eventually, become pandemic viruses. Although these mechanisms are now understood, the time frame required for adaptation of the avian virus to its new host remains unknown. Maybe the next pandemic will show us how rapid this adaptation can be.

12.1 Introduction

The world has seen pandemics of influenza A for a very long time (Creighton 1965; Major 1945; Shope 1958). Nevertheless, even if records of past epidemics describe diseases with clinical presentation resembling influenza infections, it is difficult to be certain that these epidemics of the past were the consequence of emerging influenza viruses. Recent knowledge has revealed that pandemic viruses emerge from the avian reservoir (Scholtissek 1994). In some cases, the emerging influenza requires adaptation before its dissemination in man. This adaptation can be

Bruno Lina
Laboratoire de Virologie Est, Institut de Microbiologie, Centre National de Références des virus influenza région sud, CBPE, GHE, Boulevard Pinel, 69677 Bron cedex, France;Virologie et Pathologie Humaine, UCBL-CNRS FRE3011, Université de Lyon, Faculté de Médecine R.T.H. Laennec, rue G. Paradin, 69372 Lyon, cedex 08, France
E-mail: bruno.lina@univ-lyon1.fr

D. Raoult and M. Drancourt (eds.), *Paleomicrobiology: Past Human Infections.*

achieved in an intermediate host, e.g. pigs or possibly poultry (Russell and Webster 2005). The history of pandemics seems to have started very early in mankind; some assume that influenza viruses could have become zoonotic and subsequently adapted to man when poultry started to be raised for food.

In the Sixth Book of the Epidemics, Hypocratus describes a contagious upper respiratory tract infection whose symptoms suggest an influenza-like illness. Indeed, epidemic rheumy fever and influenza-like symptoms are reported throughout human history, suggesting very contagious pathogens and very high attack rates (Creighton 1965; Patterson 1986). This conjunction is highly suggestive of influenza pandemics, especially when the mortality rates described are quite high.

This chapter describes some of the information available from past centuries about widespread upper respiratory tract infections that could have been influenza pandemics.

12.2 Before 1500

Early information on putative pandemics is difficult to collect. The first report seems to be that of Hypocratus, who described, in the Book of the Epidemics, a highly contagious disease observed in northern Greece (ca. 410 B.C.). The symptoms described very much resemble those of influenza. The second convincing report comes from England in 664 A.D. Monks reported that an epidemic swept through Britain, supposedly facilitated by clerics travelling from a synod held at Whitby Abbey (Creighton 1965).

Later reports from different parts of Europe suggest that a pandemic was observed in England, France and Italy in 1173–1174. At this time, the word "plague" was used for every epidemic responsible for significant mortality (Major 1945). A French chronicle reported that: "In May, an inflammatory plague spread all over the Occident, and all eyes swept following a cruel rhinorhea". Similar reports were provided by the churchmen of Melrose Abbey, who described "an eveil and unheard-of-cough". This might well have been an influenza pandemic, this assumption being supported by the emergence of obviously similar diseases, with very high attack rates, even mortality, at a time when the average lifespan was not longer than 30 years (Creighton 1965).

In 1357, an epidemic was described in Florence in Italy, for which the word "influenza" was used for the first time (for "influenza di freddo" or cold influence). Again, in the fourteenth century, in the South of France, a doctor from Montpellier reported an epidemic of upper respiratory tract infection that was "so important that only one out of ten of the population could escape the disease". Elderly people died in huge numbers during this epidemic, although the word "elderly" should be understood within the criteria of the times (Major 1945).

In 1414, French chroniclers described a very large epidemic that started in February. According to their writings, this epidemic was brought by a "smelly and cold wind" (in French, "vent puant et tout plein de froidure"). They reported that everyone, including clericals, nobles, and ordinary people, became infected. In Paris, up to 100,000 people were so ill that they "lost the eating and drinking. They could do nothing else than resting, they had a very high fever with shivering and cough…". This cough was so severe that the chroniclers reported numerous inguinal hernias.

During this period, different names were given to these epidemics. The 1414 epidemic was called either "tac" suggesting rapid onset, or "horizon".

Another epidemic reported in 1427 was called "dando" (sounds like "in the back" in French), and later on "coqueluche" (from cucullus meaning capuchon or small cap). This name was given to describe patients who wore coats and a cap while infected. Reports of this epidemic are available in France and England, and this epidemic is currently accepted as a true influenza pandemic. Again, clerics from St. Albans Abbey described the disease thoroughly. According to their notes, it started in February and "it invaded the whole people, and so infected the aged along with the younger that it conducted a great number to the grave." The death toll at that time was also very high in Paris. However, no records of the number of deaths are available (Creighton 1965).

12.3 Between 1500 and 1888

After 1500, descriptions of epidemics are very consistent. The first well described epidemic was reported in 1580 (Neustadt and Fineburg 1983). It clearly came from Asia during the summer, spread to Africa and then to Europe. European countries were infected northwards over a period of 6 months. Eventually, the epidemic was observed in the Americas. Numerous deaths were reported from Spanish, Italian and French cities although no figures are available. The name Influenza has been used since then to describe these massive epidemics.

The following epidemics observed in Europe are thought to have moved westwards, some crossing the Atlantic Ocean. Reports from European countries suggest that massive influenza epidemics occurred in 1658, 1679, 1708, and 1729. The latter started in Russia and had three waves, the third of which being the most severe (Patterson 1986). King Louis the XVth was infected and he called this infection a disease that spread like a foolish little girl ("follette" in French).

In 1768, Voltaire, who was in Saint Petersburg, described in a letter a disease (called "grippe" in French) that during his trip around the world passed through Siberia and infected his old body. The word "grippe" is thought to have come from a German word "grippen" that means "to catch". The word influenza, which originated in Florence, was widely used, and is seen in British reports of influenza-like diseases since 1743. At that time, it was suggested that astronomy, i.e. the conjunctions of stars and planets, could influence health status (influenza di stelle), together with cold (influenza di freddo). In 1767, in a letter to his son, Lord

Chesterfield described an epidemic in London that was very likely a seasonal epidemic. He described the disease as "a little fever that kills nobody but elderly people that is now called by a beautiful name: influenza".

In 1775, reports from a French doctor called Bachaumont describe the dissemination in France of an "epidemic common cold" that started in London. According to his writings, "…this disease is causing serious concerns to the people of Britain. Numerous are coming into southern France to escape from the disease. Since then, this plague has spread to southern France, causing numerous deaths in Toulon, Marseille and Paris". The infection rate was very high but only elderly people died. During this epidemic, up to 12 deaths of elderly people were registered per day in Paris (De Lacey 1993; Patterson 1986).

A pandemic was also certainly observed in 1781. This pandemic reached the whole world. It originated in China, spread to Russia and was subsequently disseminated westwards across Europe. It reached the east coast of Northern America in the spring of 1781 and moved westwards. This worldwide pandemic was responsible for the deaths of many young people. In Saint Petersburg, up to 30,000 cases were recorded on the same day. Similarly, it was described that half the population of Rome was infected. It is not thought to have been a pandemic with a high death rate.

Less than a decade later, in the winter of 1789–1790, influenza was widespread throughout Europe and northern America, George Washington being seriously ill (Patterson 1986).

In 1803, a large and severe epidemic was reported in France, being responsible for numerous deaths amongst the "vigorous people". Johann Freidriech Reichardt, a German ambassador in France at that time, reported that a large epidemic was observed. He described cases that could be observed from numerous places, including a large number of young people.

Other epidemics are reported in the nineteenth century (1817, 1830 and 1837), the latest of which being responsible for a very large number of cases. In 1837, a French chronicler reporter that "… half of the population of Paris was in bed, this epidemic was transforming Paris into a giant Hospital where half of the inhabitants were infected by influenza and the other half was taking care of the cases". This is a very good description of what could have been an influenza pandemic.

In 1889, a new epidemic emerged from Russia. It is said that approximately 40% of the world's population was infected, and that the influenza disease was really very severe. This pandemic has been well documented (Enserink 2006). Based on serological testing, some assume that the virus was an H2N2 subtype; however, it remains difficult to be sure of the subtype. Already at this time, some bacteriologists detected bacteria in the sputum of patients (Patterson 1986). Dr. Pfeiffer reported that unknown bacteria could be detected from the sickest of his patients. He called this bacteria Pfeiffer's bacillus (*Haemophilus influenza*) and, for a very long time, numerous microbiologists were convinced that this bacterium was the causative agent of influenza.

In 1900, a medium-sized epidemic was observed. Again, according to serological data, it is possible that this was a pandemic due to the emergence of an H3N8 strain, which was responsible for a "mild" pandemic. However, the viral subtype for both the 1889 and 1900 supposed pandemics cannot be identified with certainty. We should consider this as speculation, even if some archeoserological data support these as possible pandemics (Enserink 2006).

12.4 Virologically Confirmed Pandemics of the Twentieth Century

It is very difficult to be sure of the influenza subtypes responsible for pandemics before 1918 because there is a critical lack of specimens that can be tested for the presence of RNA that would provide consistent information regarding viral subtypes.

The first influenza viruses to be cultivated *in vitro* were isolated in 1931 from swine and in 1933 from a human specimen (Shope 1931; Smith 1935). One of these early historical strains [A/Puerto Rico/8/1934 (H1N1)] is still used for vaccine production. This virus was the circulating strain of 1934, being a variant of the H1N1 pandemic virus that emerged 16 years earlier.

The recent development of new technologies like reverse-transcription-polymerase chain reaction (RT-PCR) and reverse genetics has allowed several research teams to amplify and subsequently reconstruct viruses from pathological specimens from cases that died from influenza during the 1918 pandemic (Taubenberger et al. 1997; Taubenberger 2003). With the help of these techniques we have now identified the "original" viruses of the three pandemics of the twentieth century, and much has been learned about the mechanisms of emergence of these viruses.

12.4.1 The Spanish Flu of 1918 (A H1N1)

The so-called "Spanish Flu" has been the most devastating disease of modern times. The global death toll is estimated at 40–50 million, while 500 million to 1 billion people (representing approximately 30–50% of the world's population) are thought to have been infected (Niall et al. 2002). The history of this pandemic is well described as numerous reports are available, especially in military archives. However, the beginning of the pandemic remains obscure as yet (Reid and Taubenberger 2003). It is clear that the emergence and subsequent adaptation of the deadly A H1N1 virus took several months or even years before the start of the outbreak. This virus hit the whole world very badly within a very short time. In the early stages of the pandemic, in 1916, a French report describes a small-scale epidemic with a very high infection rate in a medium-sized city in the south of

France, with very few fatalities. Similar cases were observed in the troops of both sides during the first World War (WWI), but due to the embargo on news because of the war, there was no dissemination of this information.

Two places are suspected of being the site of emergence of the actual pandemic virus. The first hypothesis involves the province of Canton in China. The hypothesis is that this virus originated from China and subsequently travelled to the United States due to the massive immigration of Chinese people into North America (Reid and Taubenberger 2003). The virus then swept through the United States before spreading to the rest of the world (Iezzoni 1999). The second possibility is that the virus originated from the United States directly. The first cases were recorded in March 1918, and the first epidemic clusters described were located in a military camp in Furston (now called Fort Riley in Kansas), in Detroit, and in a prison in South Carolina (Soper 1918). Subsequently, the virus spread over a large part of the United States. Meanwhile, military troops sent to Europe landed in France in Brest and Bordeaux. These boats were loaded with infected soldiers. In some cases, numerous deaths were recorded during the journey across the Atlantic Ocean. The virus was introduced into Europe and cases were subsequently recorded in the French and British armies. The virus then spread to Spain, Italy, Germany and Russia. North Africa was hit in June 1918, and cases were also recorded in India, Asia and New Zealand. Until June, the pandemic was significant but no worse than previous pandemics (Niall et al. 2002). A limited number of fatal cases were recorded, mainly amongst young children and the elderly. This was only the first wave.

The second wave began at the end of August in Europe and North America. In Europe, it began in Brest, France, and in the United States in Boston. In Boston, a camp (Camp Deven) was opened to prepare the troops for war. Up to 45,000 troops were in the camp. The first case observed in this camp was recorded on the 7th of September. The following day, dozens of cases were recorded and by the 18th of September, 6,600 cases of influenza infection were recognised (Wooley 1919). At the peak of the epidemic in this camp, up to 1,176 cases were admitted to hospital in a single day. The epidemic was described as being a consequence of the dissemination of a bacterium described by Pfeiffer in 1889.

In Brest, the mortality rate was enormous, ten-fold higher than that observed during the first wave, the difference being that cases and deaths were now observed in young people (15–35 years old).

The exact spread of the epidemic remains unclear. Some suggest that the second wave originated in France and was sent back to northern America via naval ships. In North America, the major port of embarkation of the troops was Philadelphia. This city was at the origin of viral spread for the second wave in North America. The epidemic originated from the navy camp in the harbour, disseminated to the civilian population, first slowly and then rapidly, and subsequently moved westwards (Soper 1918). In North America, the second wave ended by December 1918 (Iezzoni 1999).

In Europe, dissemination of the virus was observed simultaneously. Despite the news embargo, the troops knew that a disease was responsible for a large epidemic and some tried to escape from the battlefields by describing the infection. This led to massive gatherings of infected and non-infected troops, thus boosting the epidemic.

Spain was also hit by the pandemic. This country was not at war and information was freely available. The Spanish newspapers openly described the epidemic. The name "Spanish flu" is a result of this. Moreover, the King of Spain and his court were severely hit by the virus.

During this second wave, the disease was really very severe. Although mortality was due not only to viral infections (50% was supposed to be due to bacterial super-infections), numerous cases of fulminant flu were reported. As an example, the French poet Guillaume Apollinaire fell ill on the 8th of November and died on the 9th. This rapid and deadly evolution of flu was observed also for Edmond Rostand, Gustav Klimt and Egon Schiele.

In October 1918, the number of cases in troops on both sides was so large that war was no longer possible. On the battleground, up to 37,000 United States troops and 25,000 French soldiers were ill. Historians assume that the pandemic was certainly responsible for the premature end of WWI (Crosby 1976; Patterson 1986).

At the same time, in France, the Ministry of Interior Affairs decided to close all public places, and collections of garbage were organised. Similarly, disinfectants were sprayed in places with high incidence.

The pandemic was devastating the world over. As an example, in Spitzberg, several villages were completely destroyed and the entire population died. The pandemic first hit the young. These young people died and subsequently there was nobody to hunt and look for food. As a consequence, the remaining inhabitants starved to death.

In January 1919, the third wave hit the planet. This last wave of the Spanish Flu was of lesser magnitude and ended in the spring of 1919. The impact of this last wave was very important in Australia. As an island, the Australians had decided to stop all contact with the rest of the world in an attempt to escape the epidemic. However, the virus entered Australia in 1919 with devastating results; the mortality rate was even higher than in countries that had experienced the two previous waves.

Overall, the lethality was estimated at 3.5% and the estimated number of deaths ranges from 40 million to up to 100 million (Frost et al. 1930; Niall et al. 2002).

How did this virus emerge and why was it so lethal? To the first question, there is no answer as yet. According to data collected from several teams, including that of J. Taubenberger, the A H1N1 virus that has been reconstructed shows similarities with avian viruses (Taubenberger 2003). These similarities are observed at the level of the viral nucleic acid sequences. However, analysis of the proteins shows that this virus has signatures of mammalian influenza. It is still not clear if this virus was transmitted directly from birds to man of if an intermediate host was involved. There is no record of any epidemic in poultry prior to the human pandemic. One of the differences between avian and human viruses is that they bind to different receptors (Gamblin et al. 2004; Glaser et al. 2005). Hence, a key element of the adaptation of an influenza virus to its new host is the acquisition of mutations that allow binding to the human cellular receptor. Sequences of the A H1N1 virus detected in material collected from cases that died of influenza in 1918 show a receptor binding site intermediate between avian and human, as if the virus was developing mutations to adapt to its new host (Gamblin et al. 2004). However, we do not know if the virus emerged directly from birds.

12.4.2 The Asian Flu of 1957 (A H2N2)

The second pandemic of the twentieth century was observed 40 years after the Spanish Flu. Again, this virus was thought to have emerged from China, in the province of Kweichow (Bull Org Mond Santé 1959). First reported in February 1957, this virus spread to Yunan province and moved rapidly through China. Up to 500,000 Chinese people were infected at this time. In March 1957, Mongolia and Hong Kong were hit, followed by Singapore in April. All Asia was infected by mid-May, and up to 2,000 cases were reported daily in Manilla for example. As a result of this local dissemination, the influenza strain subsequently called A2 by virologists was nicknamed "Asian flu" by the public (Bull Org Mond Santé 1959). The World Health Organisation (WHO) and virologists rapidly identified this virus as being a new strain, significantly different from the previously circulating virus known as A1 (Bull Org Mond Santé 1959). The modern classification of strains, using both surface glycoproteins for virus classification, was implemented in 1970. However, in 1957, there was knowledge that influenza viruses could be in birds, and that different serotypes could be identified. Again, there was no indication of an ongoing epizootic at the onset or before the pandemic.

The spread of this virus out of Asia was helped by aircraft and ships. As an example, an American aircraft landed in Yokosuka in Japan in April 1957. It came from Hong Kong. Upon arrival in Japan, the crew was ill. Specimens were collected from the crew and A2 viruses were detected in culture. In June, the crews of United States Navy vessels coming from Asia were heavily infected and helped introduce the virus to North America. Numerous gatherings were responsible for further dissemination of the virus (conventions, boy scout jamborees, etc.; Podosin and Felton 1958). The geographic dissemination of the A H2N2 virus has been described (Bull Org Mond Santé 1959; Cox and Subbaro 2000). Within 9 months, the virus had spread to the whole planet. The impact was much lower as compared to the 1918 pandemic. The overall estimation of the number of deaths is approximately 2 million.

Again, the emergence of the virus has been only partly deciphered. We know that this virus emerged from the animal reservoir in a more complex fashion than the A H1N1 strain in 1918. The mechanism of emergence is called genetic reassortment, a mechanism linked to the structure of the influenza A genome. This virus has a segmented genome (eight gene segments). It can infect birds (avian viruses), man (human viruses) but also other hosts including pigs (Scholtissek 1994). Pigs are interesting animals in the biology of influenza viruses since they can be infected by human viruses as well as viruses of avian origin. In birds, influenza viruses can be endemic, and numerous subtypes can circulate in wild and domestic birds (Ferguson et al. 2003; Munster et al. 2007; Scholtissek 1994).

In remote villages in Asia, families usually raise domestic birds and swine in their homes. This close proximity between animals and man can favour genetic exchange between viruses of different origin. It is assumed that the A H2N2 pandemic virus resulted from a genetic exchange between a human and an avian virus, and that this genetic exchange or reassortment occurred in pigs (Scholtissek 1994). In 1984, Scholtissek suggested that a pig had been co-infected by the human A H1N1 virus and an avian A H2N2 virus. During this co-infection, several cells of the host were co-infected and the gene segments coding for the two surface proteins (H2 and N2) were substituted with the respective gene segments of the human virus, together with a third gene segment, PB1 (coding for one of the three proteins of the polymerase complex). This reassortment resulted in a new virus with a human genetic background (five gene segments coming from the H1N1 virus) and new surface glycoproteins. This virus was adapted for dissemination in man and resulted in a new pandemic virus.

Again, the delay required for this adaptation is unknown. There was no report of any epidemic in poultry, and no A H2N2 virus was detected before the emergence of the pandemic virus. Is reassortment rapid or not? We have no indication of the time frame necessary for such genetic evolution, or if it resulted from a single event, or three successive events.

As a result of the emergence of the A H2N2 virus, the A H1N1 virus disappeared; once a new virus has been introduced into a geographic area, the previously circulating virus disappears. The mechanism for this drastic change remains unknown. We can assume a very high attack rate for an emerging virus for which nobody yet has neutralising antibodies, and it can thus spread very efficiently. Virus spread can be expressed by the reproducibility factor, R_0, which is the mean value of the number of secondary cases per index case. The R_0 value varies according to the transmissibility of the virus: a highly transmissible virus will have a high R_0. To initiate an epidemic, a virus must have an $R_0 > 1$. This value is used to construct theoretical models for the putative next pandemic (Longini et al. 2005). It was determined that the R_0 value of seasonal flu is approximately 1.4. On the other hand, the R_0 value of an emerging pandemic virus like the 1918 virus was >2 or 3. Thus, such a highly transmissible virus can block any diffusion of a previously circulating subtype (Longini et al. 2005).

In 1957, vaccine production was implemented very rapidly; the WHO initiated vaccination campaigns that started a short time after the beginning of the pandemic. This certainly reduced the impact of this pandemic as compared to that of Spanish Flu. A peculiar observation was made during the H2N2 pandemic. The impact in the elderly was lower than would have been expected, especially for those older than 80. A tentative explanation is that this age-group of patients had already encountered H2N2 viruses during the Russian flu of 1889 and could recruit neutralising antibodies from their immune memory developed during the primary infection with the A H2N2 virus that emerged at the end of the previous century. This remains speculative however.

12.4.3 *The Hong Kong Flu of 1968 (A H3N2)*

The last real pandemic was called the Hong Kong Flu. It emerged in July 1968 in Hong Kong and, like the Asian flu, spread to the rest of the world within several months (Bull Org Mond Santé 1969). It was disseminated from Hong Kong to the rest of Asia, then to Russia, Europe and the Americas. Europe and North America were hit in January 1969, and the WHO identified this virus as a new subtype quite rapidly, although not immediately. This virus was called A3 and its genetic evolution was rapidly understood.

The mechanism of emergence of this virus was very similar to that of the A H2N2 virus. It resulted from a genetic reassortment between a human and an avian virus (Scholtissek 1994). The gene segments introduced were those coding for haemagglutinin (H3) and the polymerase protein PB1. Hence, one of the surface glycoprotein (named N2) was conserved. This might explain, in part, the relatively low impact of this virus in terms of mortality (estimations of mortality are approximately 0.8 million). However, infection rates were very high, and this pandemic showed two clear waves. Again, vaccines were rapidly available.

As in 1957, the mechanism of emergence is understood, but the timeframe required for such genetic exchanges between human and avian viruses remains unknown. Again, this pandemic showed similar features as compared to the A H2N2 pandemic. Firstly, the emerging virus led to the complete extinction of the previously circulating lineage. Secondly, the impact in the elderly population was unexpectedly low. This could also be a consequence of the possible A H3N8 pandemic of 1900. In vitro studies showed that patients older than 70 had neutralising antibodies against A H3N2. These antibodies were directed against H3. Again, exposure many years previously as the explanation for pre-existing antibodies in old patients remains speculative.

12.4.4 *The Russian Pseudo-Pandemic of 1977*

In virology, one of the major differences between the first and the second parts of the twentieth century is that laboratories able to detect, cultivate and store viruses were developed during the latter. As observed after the severe acute respiratory syndrome (SARS) CoV epidemic, the risk of re-emergence of this virus resides both in a possible new introduction from its animal reservoir, or its re-introduction from laboratories holding viral stocks. This is also true for influenza viruses from previous pandemics. In 1968, the emergence of A H3N2 led to the disappearance of A H2N2, the latter in turn having being responsible for the disappearance of A H1N1. In 1969, the only human virus in circulation was A H3N2.

In 1977, in the Saint Petersburg region of Russia, many young children were infected with an A H1N1 virus. This virus was infectious only in children, who had not encountered the A H1N1 virus before its disappearance in 1957. Analysis of this emerging virus revealed that it was genetically identical to a strain that had been circulating in 1954 (Kilbourne 2006).

There are two putative explanations for the re-emergence of this subtype. First, it came out of a virology laboratory. Second, the virus survived in the permafrost or in the arctic waters of Russia and has been infecting people visiting regions where this virus remained infectious for two decades. Although this latter possibility cannot be excluded (Zhang et al. 2006), the former is the most likely. This shows that when a virus is adapted to man, its emergence even in remote places in the world can lead to its worldwide dissemination.

12.5 Putative Emerging Pandemics: 1976 A swH1N1, 1997 A H5N1, 2003 A H7N7, A H5N1 and Other Alerts

We now know that the animal reservoir for influenza viruses is enormous. There are 16 different Ha and 9 different Na subtypes, thus making a very large number of putative pandemic viruses for humanity (Munster et al. 2007).

Several cases of human infection with viruses coming from the animal reservoir have been observed. In none of these cases was the virus maintained in the human population. Such introductions must be taken very seriously. In 1976, at Fort Dix, New Jersey (USA) a soldier felt ill with flu. He died very rapidly after a training march exercise. This case was investigated and an A H1N1 virus was isolated. At that time, only the A H3N2 virus was present in man. The United States authorities feared the re-emergence of a deadly H1N1 virus and initiated a very large vaccination program. The vaccine was produced and administered to 48 million United States citizens. The vaccination program was stopped because of adverse side effects, and the lack of further cases of infection with this virus. Characterisation of the virus showed that it was a swine virus, different from the previously circulating human H1N1. No additional cases have since been reported (Neustadt and Fineburg 1983).

In May 1997, a 3-year-old boy died of influenza in Hong Kong. The virus was detected but could not be identified using the regular identification process (H1 and H3 Ha typing). After several days of analysis, the virus was identified as an H5N1 strain similar to that responsible for an epidemic in domestic birds. Between May and December 1997, 18 cases were recorded in Hong Kong, of which 6 were fatal. No human-to-human transmission was observed, all cases having been exposed to infected poultry. The authorities decided to cull domestic birds and, following culling, no further human cases were seen by the end of December (Claas et al. 1998).

In February 2003, a very large epizootic was observed in domestic birds in the Netherlands. Between March and May 2003, 85 human cases infected with A H7N7 were recorded, 1 of which was fatal (Fouchier et al. 2004). A single chain of transmission was observed (Fouchier et al. 2004). Again, control of this nascent pandemic was achieved by the massive culling of birds. Overall, 30 million birds were destroyed and the situation subsequently controlled.

The A H5N1 virus emerged again in December 2003, this time with a different genetic background as compared to the 1997 strains (Li et al. 2004). Since then, the virus has been responsible for 329 cases including 201 deaths (2 October 2007). Despite massive culling, there is no control of the epizootic and viruses have been detected in several countries in Asia, Africa and Europe. This virus is a clear pandemic threat.

This history of influenza pandemics illustrates that these events are observed rarely, but regularly. Their impact is often so great that humanity remembers these deadly outbreaks. These pandemics are not a consequence of the modern world; they simply demonstrate that influenza viruses are zoonotic viruses that can be introduced into man.

Most virologists are certain that there will be pandemics in the future. However, nobody can say when these will occur. In 2007, the best candidate for a pandemic is A H5N1, but others, such as A H9N2, may also emerge (Perdue and Swayne 2005).

References

Bull Org Mond Santé (1959) 20:183–508

Bull Org Mond Santé (1969) 41 345:335–748

Claas EC, Osterhaus AD, van Beek R, De Jong JC, Rimmelzwaan GF, Senne DA, Krauss S, Shortridge KF, Webster RG (1998) Human influenza A H5N1 virus related to a highly pathogenic avian influenza virus. Lancet 351:472–477

Cox NJ, Subbaro K (2000) Global epidemiology of influenza: past and present. Annu Rev Med 51:407–421

Creighton C (1965) A history of the epidemics in Britain, vols 1, 2. Barnes and Noble, New York

Crosby AW Jr (1976) Epidemic and peace, 1918. Greenwood, Westport, CT

De Lacey M (1993) The conceptualisation of influenza in eighteenth-century Britain: specificity and contagion. Bull Hist Med 67:74–118

Enserink M (2006) What came before 1918: archaeovirologist offers a first glimpse. Science 312:1725

Ferguson NM, Galvani AP, Bush RM (2003) Ecological and immunological determinants of influenza evolution. Nature 422:428–433

Fouchier RA, Schneeberger PM, Rozendaal FW, Broekman JM, Kemink SA, Munster V, Kuiken T, Rimmelzwaan GF, Schutten M, Van Doornum GJ, Koch G, Bosman A, Koopmans M, Osterhaus AD (2004) Avian influenza A virus (H7N7) associated with human conjunctivitis and a fatal case of acute respiratory distress syndrome. Proc Natl Acad Sci USA 101:1356–1361

Gamblin SJ, Haire LF, Russell RJ, Stevens DJ, Xiao B, Ha Y, Vasisht N, Steinhauer DA, Daniels RS, Elliot A, Wiley DC, Skehel JJ (2004) The structure and binding properties of the 1918 influenza haemagglutinin. Science 303:1838–1842

Glaser L, Stevens J, Zamarin D, Wilson IA, Garcia-Sastre A, Tumpey TM, Basler CF, Taubenberger JK, Palese P (2005) A single amino acid substitution in 1918 influenza virus hemagglutinin changes receptor binding specificity. J Virol 79:11533–11566

Frost WH, Gover M, Sydenstryker E (1930) Mortality from influenza and pneumonia in 50 large cities of the United States, 1910–1929. Public Health Rep 45:2277–2328

Iezzoni L (1999) Influenza 1918. The worst epidemic in American history. TV Books, New York

Kilbourne ED (2006) Influenza pandemics of the 20th century. Emerg Infect Dis 12:9–14

Li KS, Guan Y, Smith GJD, Xu KM, Duan L, Rahardjo AP, Puthavathana P, Buranathal C, Nguyen TD, Estoepangestie ATS, Chaisingh A, Auewarakui P, Long HT, Hanh NTH Webby RJ, Poon LLM, Shortridge KF, Yuen KY, Webster RGF, Peiris JSM (2004) Genesis of a highly pathogenic and potentially pandemic H5N1 influenza virus in eastern Asia. Nature 430:209–213

Longini IM Jr, Nizam A, Xu S, Ungchuzak K, Hansaoworakul W, Cummings DA, Halloran ME (2005) Containing pandemic influenza at the source. Science 309:1083–1087

Major RH (1945) Classic descriptions of disease with biographical sketches of the authors, 3rd edn. Thomas, Springfield, IL

Munster VJ, Baas C, Lexmond P, Waldenström J, Wallensten A, Fransson T, Rimmelzwaan GF, Beyer WEP, Neustadt RE, Fineburg H (1983) The epidemic that never was. Vintage, New York

Schutten M, Olsen B, Osterhaus ADME, Fouchier RAM (2007) Spatial, temporal, and species variation in prevalence of influenza A viruses in wild migratory birds. PLoS Pathog 3:e61

Niall P, Johnson AS, Mueller J (2002) Updating the accounts: global mortality of the 1918–1920 "Spanish" influenza epidemic. Bull Hist Med 76:105–115

Patterson KD (1986) Pandemic influenza 1700–1900. Rowman and Littlefield, Totowa, NJ

Perdue ML, Swayne DE (2005) Public health risk from avian influenza viruses. Avian Dis 49:317–327

Podosin RL, Felton WL II (1958) The clinical picture of far east influenza occurring at the fourth national boy scout jamboree. Report of 616 cases. New Engl J Med 258:778–782

Reid AH, Taubenberger JK (2003) The origin of the 1918 pandemic influenza virus: a continuing enigma. J Gen Virol 84:2285–2292

Russell CJ, Webster RG (2005) The genesis of a pandemic influenza virus. Cell 123:368–371

Scholtissek C (1994) Source for Influenza pandemics. Eur J Epidemiol 10:455–458

Shope RE (1931) Swine influenza III. Filtration experiments and aetiology. J Exp Med 54:373–385

Shope RE (1958) Influenza. History, epidemiology, and speculation. Public Health Rep 73:165–178

Smith W (1935) Cultivation of the virus of influenza. Br J Exp Pathol 16:508–512

Soper GA (1918) The pandemic in army camps. JAMA 71:1899–1909

Taubenberger JK (2003) Genetic characterization of the 1918 "Spanish" influenza virus. In: Phillips H, Taubenberger JK, Reid AH, Krafft AE, Bijwaard KE, Fanning TG (1997) Initial genetic characterization of the 1918 "Spanish" influenza virus. Science 275:1793–1796

Killingray D (eds) The Spanish Influenza pandemic of 1918–19. New perspectives. Routeledge, London, pp 36–46

Wooley PG (1919) The epidemic of influenza in Camp Devens, Mass. J Lab Clin Med 4:330–343

Zhang G, Shoham D, Gilichinsky D, Davydov S, Castello JD, Rogers SO (2007) Evidence of influenza A virus RNA in Siberian lake ice. J Virol 2006 80:12229–12235 (Erratum in J Virol 81:2538)

Parasites

Chapter 13
Human lice: Pediculus and Pthirus

Kosta Y. Mumcuoglu

Abstract Lice have probably been associated with humans since the times of our pre-hominid ancestors, and were dispersed throughout the world by early human migrants. It has been suggested that the head louse is the ancestor of the human louse, and that the body louse developed later when hominids started to wear clothing. Lice are mentioned in the Bible as the third plague. From Sumerian, Akkadian, and Egyptian sources it is also evident that the ancient inhabitants of the Middle East were well acquainted with head lice. Head lice and eggs have been found on the hair of Egyptian mummies. Nine-thousand-year-old louse eggs were found in hair samples from an individual who lived in a cave near the Dead Sea in Israel, while large numbers of lice were recovered from a 3,800-year-old female mummy from the Loulan period. Louse combs from Pharonic times in Egypt were used for delousing. Head lice and their eggs have also been found in combs recovered from archaeological excavations in the Judean and Negev deserts of Israel, including from Masada and Qumran. Body lice eggs have been found in pre-historic textiles from Austria; this louse was also recovered from deposits of farmers in Viking Greenland. The remains of a body louse were also found in one of the rooms at the Masada fortress dating from the Roman period. The oldest known pubic lice are from the Roman period in Britain and from post-medieval deposits in Iceland.

13.1 Human Lice

The human louse, *Pediculus humanus*, is probably one of the oldest ectoparasites of man (Zinsser 1935). Humans are parasitised by two sub-species: the head louse *Pediculus humanus capitis*, and the body louse *Pediculus humanus humanus*. A close relative of this species, *Pediculus mjobergi*, is a parasite of South American monkeys of the family Cebidae (Retanda Salazar 1994)

Kosta Y. Mumcuoglu
Department of Parasitology, Hebrew University-Hadassah Medical School, P.O.B. 12272, Jerusalem 91120, Israel E-mail: kostam@cc.huji.ac.il

D. Raoult and M. Drancourt (eds.), *Paleomicrobiology: Past Human Infections.*

Lice have probably been associated with humans since the time of our prehominid ancestors, and were dispersed throughout the world by early human migrants (Marsh 1964). It has been suggested that the head louse is the ancestor of the human louse, and that the body louse developed later when hominids started to wear clothing (Maunder 1983).

Reed et al. (2004) proposed an evolutionary history of *P. humanus* based on morphological and genetic analyses, and confirmed that *P. humanus* has two lineages – one comprising the head and body forms with worldwide distribution, and the other consisting of the head louse restricted only to the New World. They came to the conclusion that *P. humanus* originated long before its human host.

Humans went through a population bottleneck around 100,000 years ago, followed by expansion. Population genetics studies of human lice revealed that only the worldwide lineage passed through this bottleneck and subsequent expansion. The New World lineage has not only maintained a relatively stable population size but has followed an evolutionary path distinct from that of the worldwide lineage for the past 1.2 million years. It has also been suggested that these two ancient louse lineages could have embarked on these different evolutionary pathways on a single host. More likely, the New World louse evolved on an archaic form of human before casting its lot with a modern version. While the split between *Homo sapiens* and *Homo neanderthalensis* was too recent (about 700,000 years ago) to support a concurrent split between the worldwide and New World lice lineages, the split between *H. sapiens* and *H. erectus* (about 1.8 million years ago) could have. Reed and colleagues (2004) proposed a scenario in which *H. sapiens* and *H. erectus* carried distinct types of lice owing to approximately 1 million years of separation. As the first waves of modern humans left Africa about 100,000 years ago and modern humans replaced archaic forms, the two forms engaged in enough contact for archaic lice to make the switch to modern human hosts.

Lice are mentioned in the Bible as the third plague visited on the Egyptians when the Pharaoh denied the request of Moses to let the Israelites go. From Sumerian, Akkadian, Egyptian, and Biblical sources it is evident that the ancient inhabitants of the Middle East were well acquainted with head lice (Bodenheimer 1947/1948; Driver 1974; Aufderheide and Rodriguez-Martin 1998). In the sixteenth century B.C., an Egyptian text, known as the Papyrus Ebers, described a remedy for lice prepared from date flour.

In the Near East, head lice and eggs have been found on the hair of Egyptian mummies (Ruffer 1921; Hoeppli 1956; Fletcher 1994). Nine-thousand-year-old louse eggs were found on hair samples from an individual who lived in Nahal Hemar Cave near the Dead Sea in Israel (Mumcuoglu and Zias 1991).

In Asia, large numbers of lice were recovered from a 3,800-year-old female mummy from the Loulan period (Wen et al. 1987).

In Europe, ancient head lice are known from the Roman period onwards (Hall and Kenward 1990; Schelvis 1994; Kenward and Hall 1995; Allison et al. 1999), and there are also records from Iceland (Amorosi et al. 1992; Buckland et al. 1992) and Greenland (Buckland et al. 1983; Bresciani et al. 1983; Hansen and Gullov 1989; Sadler 1990).

In North America, head lice and their eggs have been found on mummified remains of prehistoric Indians from the American Southwest (Ewing 1924; Graham 1965; Horne 1979; Cockburn and Cockburn 1980; Cockburn 1983). Lice have been found in hunter-gatherer and agricultural sites in the Unites States (the Great Basin of Utah and surrounding states, and the Colorado Plateau) and in central Mexico (El-Najjar 1998). The prehistoric peoples in these areas appeared to control the lice by eating lice groomed from hair (a common method of louse control among tribal cultures, even today) as adult lice have been found deep in the matrix of coprolites (Fry 1977; Reinhard et al. 1986; Reinhard and Largent 1989; Reinhard 1990).

In South America, lice were found on the mummy of an Inca prince, who lived approximately 500 years ago (Horne and Kawasaki 1984) as well as on mummified pre-Columbian Indians from Peru (Fletcher 1994; Reinhard and Buikstra 2003). Head louse eggs were recovered from human hair found in Brazil and were carbon dated to approximately 10,000 years old (Araujo et al. 2000). Hair samples from seven mummies from Camarones, Chile, carbon-dated to ca. 1900–1500 B.C., were examined and head lice eggs were found in six of them (M.A. Rivera, K.Y. Mumcuoglu, R.T. Matheny and D.G. Matheny, manuscript submitted) (Fig. 13.1).

The oldest combs similar to today's louse combs date from 1500 B.C. (Zias and Mumcuoglu 1989). Royal combs from Pharonic times in Egypt were used for delousing (Kamal 1967). Head lice were recovered from the debris found between the fine teeth of a wooden comb excavated in Antionoe, Egypt and dated between the fifth and sixth centuries A.D. (Palma 1991).

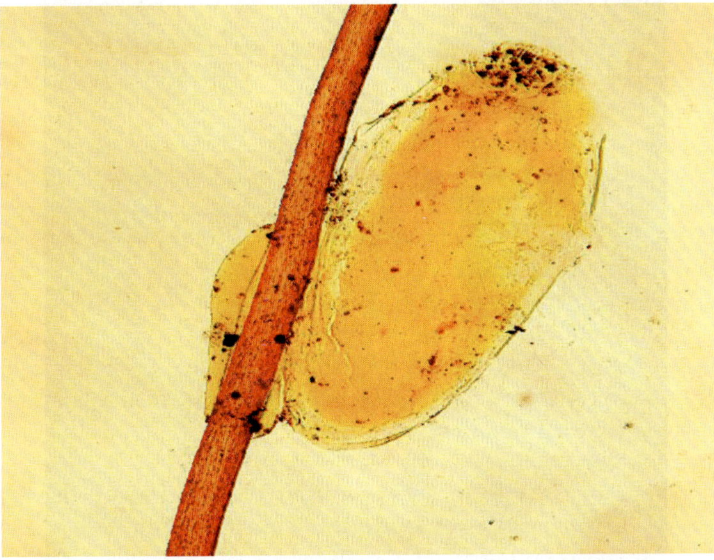

Fig. 13.1 Operculated egg found on the scalp of a mummy from the Chinchorro Tradition, Camarones, Northern Chile

Head lice and their eggs have also been found in combs recovered from archaeological excavations in the Judean and Negev deserts of Israel, including from Masada and Qumran (Fig. 13.2). Most of the combs were two-sided (Fig. 13.3), while some were also single-sided (Fig. 13.4). One side of the comb was used to open the knots while the second side with the fine teeth was used to remove lice and eggs. Most combs found in archaeological excavations were made out of wood; some were made from bones and ivory, yet all bear a resemblance to modern day combs. Lice were found in 12 out of 24 combs examined from the Judean and Negev Deserts. In a comb from Wadi Farah, 4 lice and 88 eggs were found; 2 of them were operculated, showing that at this stage the eggs were viable with an embryo inside. In one comb from Qumran, 12 lice and 27 eggs were found, 10 of them operculated (Mumcuoglu and Zias 1988).

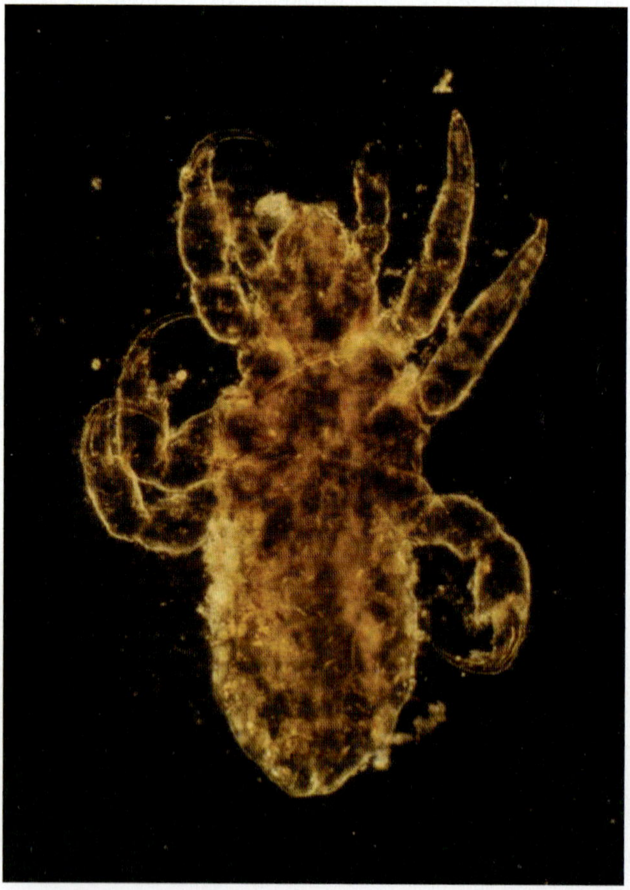

Fig. 13.2 Second nymphal stage of a head louse from a comb from Qumran, Israel (68 A.D.)

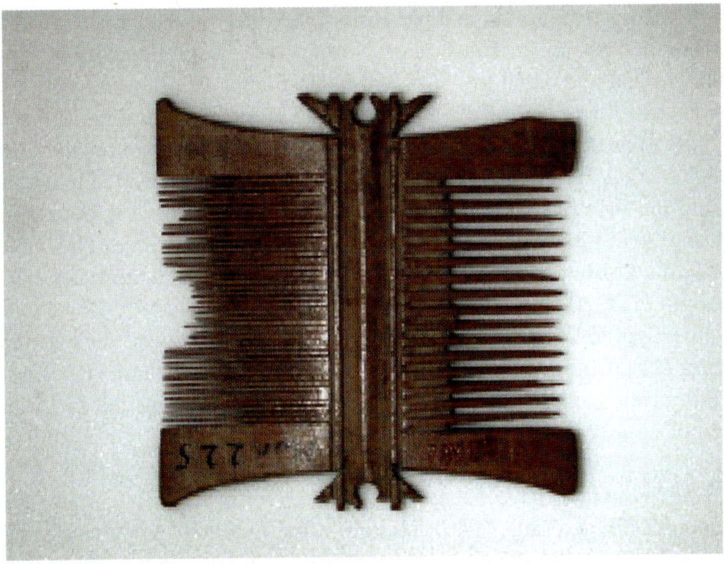

Fig. 13.3 Two-sided wooden comb from the Judean desert, Israel (135 A.D.)

Fig. 13.4 Single-sided wooden comb from the Jordan Valley in Israel (eighth century A.D.)

Body lice eggs were found in a pre-historic textile from Hallstaetter Salzberg in Austria (Hundt 1960). This louse was also recovered from deposits of farmers in Viking Greenland and dated to 986–1350 A.D. (Sadler 1990).

The remains of a body louse were also found in one of the rooms at the Masada fortress known as the "Casemate of the Scrolls". Originally constructed during the last decade of King Herod's reign, the Casemate Room was converted into a dwelling unit during the first Jewish revolt against the Romans. Following the conquest of Masada, the room was used by Roman soldiers as a dumping area. The context and nature of the textiles associated with the louse clearly suggest a rebel origin (Mumcuoglu et al. 2003).

The oldest pubic lice (*Pthirus pubis*) found in archaeological deposits are from the Roman period (mid-first or second centuries A.D.) in Britain (Buckland et al. 1992). Pubic lice have also been found in post-medieval deposits in Iceland and from samples collected from archaeological remains from eighteenth century London (Girling 1984; Kenward 1999, 2001). There are early Chinese, Greek and Roman sources, which have been interpreted as referring to pubic lice (Busvine 1976; Hoeppli and Chi'ang 1940), including the treatment of infestation of eyelashes, which, although rare, also occurs in present times (Burns 1987).

Thirty-seven mummies from San Pedro de Atacama, dated up to 2,000 years old, were examined for parasites. Pubic hair was present in four mummies, and eggs were found attached to the pubic hair in one adult male mummy. Specimens of this parasite were also found on the pubic hair of a mummy from Chiribaya Bajan (Peru), which was dated to 1050–800 B.C., and in the pleats of a piece of cloth associated with a female mummy (Rick et al. 2002).

References

Allison EP, Kenward HK, Tomlinson PR (1990) Environmental analysis of samples from Area 86.53H, Peel Castle, Isle of Man. Environmental Archeology Unit Report 90/7, York

Amorosi T, Buckland PC, Olafsson G, Sadler JP Skidmore P (1992) Site status and the palaeoecological record: a discussion of the results from Bessastadir, Iceland. In: Morris CD, Rackham JD (eds) Norse and later settlement and subsistence in the North Atlantic. Department of Archaeology, University of Glasgow, pp 169–192

Araujo A, Ferreira LF, Guidon N, Maues de Serra Freire N, Reinhard KJ, Dittmar K (2000) Ten thousand years of head lice infection. Parasitol Today 16:269

Aufderheide A, Rodriguez-Martin C (1998) Cambridge encyclopedia of human paleopathology. Cambridge University Press, Cambridge

Bodenheimer FS (1947/1948) A survey on the ancient Sumerians and Assyrians. Arch Int Hist Sci 1:261–269

Bresciani J, Haarlov N, Nansen P, Moller G (1983) Head louse *Pediculus humanus capitis* from mummified corpses of Greenlanders about A.D. 1460. Act Entomol Fenn 42:24–27

Buckland PC, Sveinbjarnardottir G, Savory D, McGovern TH, Skidmore P, Andreasen C (1983) Norseman at Nipaitsoq, Greenland: a palaeoecological investigation. Norw Archeol Rev 16:86–98

Buckland PC, Sadler JP, Sveinbjarnardottir G (1992) Palaeoecological Investigations at Reykholt, Western Iceland. In: Morris CD, Rackham DJ (eds) Norse and Later Settlement and

Subsistence in the North Atlantic. Department of Archaeology, University of Glasgow, pp 149–168
Burns DA (1987) The treatment of *Phthirus pubis* infestation of the eyelashes. Br J Dermatol 117:741–743
Busvine JR (1976) Insects, hygiene and history. Althone, London
Cockburn A (1983) Mummies, disease and ancient cultures, abridged edn. Cambridge University Press, Cambridge
Cockburn A, Cockburn E (1980) Mummies, disease and ancient culture, 1st edn. Cambridge University Press, Cambridge
Driver GC (1974) Lice in the Old Testament. Palest Explor Q 106:159–160
El-Najjar MY (1998) Mummies and mummification practices in the southwestern and southeastern United States. In: Cockburn A, Cockburn E, Reyman T (eds) Mummies, disease, and ancient cultures, 2nd edn. Cambridge University Press, Cambridge, pp 121–137
Ewing HE (1924) Lice from human mummies. Science 60:389–390
Fletcher I (1994) A tale of hair, wigs and lice. Egypt Archaeol 5:31–33
Fry GF (1977) Analysis of prehistoric coprolites from Utah. University of Utah Anthropological Papers 9
Girling MA (1984) Eighteenth century records of human lice (Phthiraptera, Anoplura) and fleas (Siphonaptera, Pulicidae) in the City of London. Entomol Mon Mag 120:207–210
Graham SA (1965) Entomology: an aid in archaeological studies. Am Antiq Mem 19:167–174
Hall AR, Kenward HK (1990) Environmental evidence from the Colonia: General Accident and Rougier Street. In: Addyman PV (ed) The Archaeology of York. CBA, London, pp 289–434
Hansen JPH, Gullov HC (1989) The mummies from Qilakilsoq – paleopathological aspects. Meddr Grønland Man Soc 12:69–82
Hoeppli R, Ch'iang IH (1940) The louse, crab louse and bed-bug in old Chinese medical literature, with special consideration on phthiriasis. Chin Med J 58:338–362
Horne P (1979) Head lice from an Aleutian mummy. Paleopathol News 25:7–8
Horne PD, Kawasaki SQ (1984) The prince of El Plomo. A paleopathological study. Bull N Y Acad Med 60:925–931
Hoeppli R (1956) The knowledge of parasites and parasite infections from ancient times to the 17th century. Exp Parasitol 5:398–412
Hundt HJ (1960) Vorgeschichtliche Gewebe aus dem Hallstaetter Salzberg. Jahrb Roemisch-Germanischen Zentralmuseums Mainz 7:126–141
Kamal H (1967) A dictionary of Pharonic medicine. National Publication House, Cairo
Kenward H (1999) Pubic lice (*Pthirus pubis* L.) were present in Roman and Medieval Britain. Antiquity 73:911–915
Kenward H (2001) Pubic lice in Roman and medieval Britain. Trends Parasitol 17:167–168
Kenward HK, Hall AR (1995) Biological evidence from Anglo-Scandinavian deposits at 16–22 Coppergate. In: In: Addyman PV (ed) The Archaeology of York. CBA, London, pp 435–797
Marsh DC (1964) Pre-Columbian insects from Tamaulipas, Mexico. Phytoprotection 45:134
Maunder JW (1983) The appreciation of lice. Proc R Inst G B 5:1–31
Mumcuoglu KY, Zias J (1988) Head lice, *Pediculus humanus capitis* (Anoplura: Pediculidae) from hair combs excavated in Israel and dated from the first century B.C. to the eighth century A.D. J Med Entomol 25:545–547
Mumcuoglu KY, Zias J (1991) Pre-pottery neolithic B head lice found in Nahal Hemar Cave and dated 6,900–6,300 B.C.E. (uncalibrated). Atikot 20:167–168
Mumcuoglu KY, Zias J, Tarshis M, Lavi M, Stiebel GD (2003) Body louse remains in textiles excavated at Massada, Israel. J Med Entomol 40:585–587
Palma RL (1991) Ancient head lice on a wooden comb from Antinoe, Egypt. J Egypt Archaeol 77:194
Reed DL, Smith VS, Hammond SL, Rogers AR, Clayton DH (2004) Genetic analysis of lice supports direct contact between modern and archaic humans. PLoS Biol 2:e340, DOI: 10.1371/journal.pbio.0020340

Reinhard K (1990) Archaeoparasitology in North America. Am J Phys Anthropol 82:145–163
Reinhard KJ, Buikstra J (2003) Louse infestation of the Chiribaya culture, southern Peru: variation in prevalence by age and sex. Mem Inst Oswaldo Cruz 98:173–179
Reinhard KJ, Largent FB (1989) Diet and environment determined from analysis of prehistoric coprolites from an archaeological site near Zape Chico, Durango, Mexico. J Paleopathol 1:151–156
Reinhard K, Confalonieri U, Herrmann B, Ferreira LF, Araujo A (1986) Recovery of parasite remains from coprolites and latrines: aspects of paleoparasitological technique. Homo 37:217–239
Retanda Salazar AP (1994) Filogenia de los piojos (Insecta: Anoplura) de los monos del Viejo Mundo (Catarrhini). Rev Biol Trop 42:633–638
Rick FM, Rocha GC, Dittmar K, Coimbra CEA, Reinhard K, Bouchet F, Ferreira LF, Araujo A (2002) Crab louse infestation in Pre-Columbian America. J Parasitol 88:1266–1267
Ruffer MA (1921) Studies in the paleopathology of Egypt. University of Chicago Press, Chicago, IL
Sadler JP (1990) Records of ectoparasites on humans and sheep from Viking-Age deposits in the former western settlement of Greenland. J Med Entomol 27:628–631
Schelvis J (1994) Caught between the teeth. A review of Dutch finds of archeological remains of ectoparasites in combs. Proceedings of the Section Experimental and Applied Entomology of the Netherlands Entomological Society 5:131–132
Wen T, Zhaoyong X, Zhijie G, Yehua X, Jianghua S, Zhiyi G (1987) Observation on the ancient lice from Loulan. Investigatio et Studium Naturae (Museum Historiae Naturae, Shanghaiense) 7:152–155
Zias J, Mumcuoglu KY (1989) How the ancients de-loused themselves. Bibl Archaeol Rev 15:66–69
Zinsser H (1935) Rats, lice and history. Little Brown, Boston

Index

M

N

O

P

Q

R